# Disclaimer

The publisher of this book is by no way associated with the National Institute of Standards and Technology (NIST). The NIST did not publish this book. It was published by 50 page publications under the public domain license.

50 Page Publications.

**Book Title:** Electronic Authentication Guideline

**Book Author:** William E. Burr; Donna F. Dodson; Elaine M. Newton; Ray A. Perlner; William T. Polk; Sarbari Gupta; Emad A. Nabbus;

**Book Abstract:** This recommendation provides technical guidelines for Federal agencies implementing electronic authentication and is not intended to constrain the development or use of standards outside of this purpose. The recommendation covers remote authentication of users (such as employees, contractors, or private individuals) interacting with government IT systems over open networks. It defines technical requirements for each of four levels of assurance in the areas of identity proofing, registration, tokens, management processes, authentication protocols and related assertions. This publication supersedes NIST SP 800-63.

**Citation:** NIST SP - 800-63-1

**Keyword:** Authentication; Authentication Assurance; Credential Service Provider; Cryptography; Electronic Authentication; Electronic Credentials; Electronic Transactions; Electronic Government; Identity Proofing; Passwords; PKI; Public Key Infrastructure; Tokens.

NIST Special Publication 800-63-1

# Electronic Authentication Guideline

*Recommendations of the National Institute of Standards and Technology*

**William E. Burr**
**Donna F. Dodson**
**Elaine M. Newton**
**Ray A. Perlner**
**W. Timothy Polk**
**Sarbari Gupta**
**Emad A. Nabbus**

# INFORMATION SECURITY

Computer Security Division
Information Technology Laboratory
National Institute of Standards and Technology
Gaithersburg, MD 20899-8930

*December 2011*

**U.S. Department of Commerce**
*John E. Bryson, Secretary*

**National Institute of Standards and Technology**
*Patrick D. Gallagher, Under Secretary for Standards and Technology and Director*

## Reports on Computer Systems Technology

The Information Technology Laboratory (ITL) at the National Institute of Standards and Technology (NIST) promotes the U.S. economy and public welfare by providing technical leadership for the nation's measurement and standards infrastructure. ITL develops tests, test methods, reference data, proof of concept implementations, and technical analyses to advance the development and productive use of information technology. ITL's responsibilities include the development of management, administrative, technical, and physical standards and guidelines for the cost-effective security and privacy of other than national security-related information in federal information systems. The Special Publication 800-series reports on ITL's research, guidelines, and outreach efforts in information system security, and its collaborative activities with industry, government, and academic organizations.

## Authority

This publication has been developed by NIST to further its statutory responsibilities under the Federal Information Security Management Act (FISMA), Public Law (P.L.) 107-347. NIST is responsible for developing information security standards and guidelines, including minimum requirements for federal information systems, but such standards and guidelines shall not apply to national security systems without the express approval of appropriate federal officials exercising policy authority over such systems. This guideline is consistent with the requirements of the Office of Management and Budget (OMB) Circular A-130, Section 8b(3), *Securing Agency Information Systems*, as analyzed in Circular A-130, Appendix IV: *Analysis of Key Sections*. Supplemental information is provided in Circular A-130, Appendix III, *Security of Federal Automated Information Resources*.

Nothing in this publication should be taken to contradict the standards and guidelines made mandatory and binding on federal agencies by the Secretary of Commerce under statutory authority. Nor should these guidelines be interpreted as altering or superseding the existing authorities of the Secretary of Commerce, Director of the OMB, or any other federal official. This publication may be used by nongovernmental organizations on a voluntary basis and is not subject to copyright in the United States. Attribution would, however, be appreciated by NIST.

NIST Special Publication 800-63-1, 121 pages

**(December 2011)**

---

Certain commercial entities, equipment, or materials may be identified in this document in order to describe an experimental procedure or concept adequately. Such identification is not intended to imply recommendation or endorsement by NIST, nor is it intended to imply that the entities, materials, or equipment are necessarily the best available for the purpose.

There may be references in this publication to other publications currently under development by NIST in accordance with its assigned statutory responsibilities. The information in this publication, including concepts and methodologies, may be used by federal agencies even before the completion of such companion publications. Thus, until each publication is completed, current requirements, guidelines, and procedures, where they exist, remain operative. For planning and transition purposes, federal agencies may wish to closely follow the development of these new publications by NIST.

Organizations are encouraged to review all draft publications during public comment periods and provide feedback to NIST, per the instructions provided for the draft document at http://csrc.nist.gov/publications/PubsDrafts.html. All NIST publications, other than the ones noted above, are available at http://csrc.nist.gov/publications.

## Abstract

This recommendation provides technical guidelines for Federal agencies implementing electronic authentication and is not intended to constrain the development or use of standards outside of this purpose. The recommendation covers remote authentication of users (such as employees, contractors, or private individuals) interacting with government IT systems over open networks. It defines technical requirements for each of four levels of assurance in the areas of identity proofing, registration, tokens, management processes, authentication protocols and related assertions. This publication supersedes NIST SP 800-63.

KEY WORDS: Authentication, Authentication Assurance, Credential Service Provider, Cryptography, Electronic Authentication, Electronic Credentials, Electronic Transactions, Electronic Government, Identity Proofing, Passwords, PKI, Public Key Infrastructure, Tokens.

## Acknowledgments

The authors, William Burr, Donna Dodson, Elaine Newton, Ray Perlner, Tim Polk of the National Institute of Standards and Technology (NIST), and Sarbari Gupta, and Emad Nabbus of Electrosoft, would like to acknowledge the contributions of our many reviewers, including Jim Fenton from Cisco, Hildegard Ferrailo from NIST, and Erika McCallister from NIST, as well as those from Enspier, Orion Security, and Mitre.

## Executive Summary

Electronic authentication (e-authentication) is the process of establishing confidence in user identities electronically presented to an information system. E-authentication presents a technical challenge when this process involves the remote authentication of individual people over an open network, for the purpose of electronic government and commerce. The guidelines in this document assume the authentication and transaction take place across an open network such as the Internet. In cases where the authentication and transaction take place over a controlled network, agencies may take these security controls into account as part of their risk assessment.

This recommendation provides technical guidelines to agencies to allow an individual to remotely authenticate his or her identity to a Federal IT system. This document may inform but does not restrict or constrain the development or use of standards for application outside of the Federal government, such as e-commerce transactions. These guidelines address only traditional, widely implemented methods for remote authentication based on secrets. With these methods, the individual to be authenticated proves that he or she knows or possesses some secret information.

Current government systems do not separate functions related to identity proofing in registration from credential issuance. In some applications, credentials (used in authentication) and attribute information (established through identity proofing) could be provided by different parties. While a simpler model is used in this document, it does not preclude agencies from separating these functions.

These technical guidelines supplement OMB guidance, *E-Authentication Guidance for Federal Agencies* [OMB M-04-04] and supersede NIST SP 800-63. OMB M-04-04 defines four levels of assurance, Levels 1 to 4, in terms of the consequences of authentication errors and misuse of credentials. Level 1 is the lowest assurance level, and Level 4 is the highest. The OMB guidance defines the required level of authentication assurance in terms of the likely consequences of an authentication error. As the consequences of an authentication error become more serious, the required level of assurance increases. The OMB guidance provides agencies with the criteria for determining the level of e-authentication assurance required for specific applications and transactions, based on the risks and their likelihood of occurrence of each application or transaction.

OMB guidance outlines a 5-step process by which agencies should meet their e-authentication assurance requirements:

1. Conduct a risk assessment of the government system.
2. Map identified risks to the appropriate assurance level.
3. Select technology based on e-authentication technical guidance.

4. Validate that the implemented system has met the required assurance level.
5. Periodically reassess the information system to determine technology refresh requirements.

This document provides guidelines for implementing the third step of the above process. After completing a risk assessment and mapping the identified risks to the required assurance level, agencies can select appropriate technology that, at a minimum, meets the technical requirements for the required level of assurance. In particular, the document states specific technical requirements for each of the four levels of assurance in the following areas:

- Identity proofing and registration of Applicants (covered in Section 5),
- Tokens (typically a cryptographic key or password) for authentication (covered in Section 6),
- Token and credential management mechanisms used to establish and maintain token and credential information (covered in Section 7),
- Protocols used to support the authentication mechanism between the Claimant and the Verifier (covered in Section 8),
- Assertion mechanisms used to communicate the results of a remote authentication if these results are sent to other parties (covered in Section 9).

A summary of the technical requirements for each of the four levels is provided below.

**Level 1** - Although there is no identity proofing requirement at this level, the authentication mechanism provides some assurance that the same Claimant who participated in previous transactions is accessing the protected transaction or data. It allows a wide range of available authentication technologies to be employed and permits the use of any of the token methods of Levels 2, 3, or 4. Successful authentication requires that the Claimant prove through a secure authentication protocol that he or she possesses and controls the token.

Plaintext passwords or secrets are not transmitted across a network at Level 1. However this level does not require cryptographic methods that block offline attacks by eavesdroppers. For example, simple password challenge-response protocols are allowed. In many cases an eavesdropper, having intercepted such a protocol exchange, will be able to find the password with a straightforward dictionary attack.

At Level 1, long-term shared authentication secrets may be revealed to Verifiers. At Level 1, assertions and assertion references require protection from manufacture/modification and reuse attacks.

**Level 2** – Level 2 provides single factor remote network authentication. At Level 2, identity proofing requirements are introduced, requiring presentation of identifying materials or information. A wide range of available authentication technologies can be employed at Level 2. For single factor authentication, Memorized Secret Tokens, Pre-Registered Knowledge Tokens, Look-up Secret Tokens, Out of Band Tokens, and Single

Factor One-Time Password Devices are allowed at Level 2. Level 2 also permits any of the token methods of Levels 3 or 4. Successful authentication requires that the Claimant prove through a secure authentication protocol that he or she controls the token. Online guessing, replay, session hijacking, and eavesdropping attacks are resisted. Protocols are also required to be at least weakly resistant to man-in-the middle attacks as defined in Section 8.2.2.

Long-term shared authentication secrets, if used, are never revealed to any other party except Verifiers operated by the Credential Service Provider (CSP); however, session (temporary) shared secrets may be provided to independent Verifiers by the CSP. In addition to Level 1 requirements, assertions are resistant to disclosure, redirection, capture and substitution attacks. Approved cryptographic techniques are required for all assertion protocols used at Level 2 and above.

**Level 3** – Level 3 provides multi-factor remote network authentication. At least two authentication factors are required. At this level, identity proofing procedures require verification of identifying materials and information. Level 3 authentication is based on proof of possession of the allowed types of tokens through a cryptographic protocol. Multi-factor Software Cryptographic Tokens are allowed at Level 3. Level 3 also permits any of the token methods of Level 4. Level 3 authentication requires cryptographic strength mechanisms that protect the primary authentication token against compromise by the protocol threats for all threats at Level 2 as well as verifier impersonation attacks. Various types of tokens may be used as described in Section 6.

Authentication requires that the Claimant prove, through a secure authentication protocol, that he or she controls the token. The Claimant unlocks the token with a password or biometric, or uses a secure multi-token authentication protocol to establish two-factor authentication (through proof of possession of a physical or software token in combination with some memorized secret knowledge). Long-term shared authentication secrets, if used, are never revealed to any party except the Claimant and Verifiers operated directly by the CSP; however, session (temporary) shared secrets may be provided to independent Verifiers by the CSP. In addition to Level 2 requirements, assertions are protected against repudiation by the Verifier.

**Level 4** – Level 4 is intended to provide the highest practical remote network authentication assurance. Level 4 authentication is based on proof of possession of a key through a cryptographic protocol. At this level, in-person identity proofing is required. Level 4 is similar to Level 3 except that only "hard" cryptographic tokens are allowed. The token is required to be a hardware cryptographic module validated at Federal Information Processing Standard (FIPS) 140-2 Level 2 or higher overall with at least FIPS 140-2 Level 3 physical security. Level 4 token requirements can be met by using the PIV authentication key of a FIPS 201 compliant Personal Identity Verification (PIV) Card.

Level 4 requires strong cryptographic authentication of all communicating parties and all sensitive data transfers between the parties. Either public key or symmetric key technology may be used. Authentication requires that the Claimant prove through a secure authentication protocol that he or she controls the token. All protocol threats at Level 3 are required to be prevented at Level 4. Protocols shall also be strongly resistant to man-in-the-middle attacks. Long-term shared authentication secrets, if used, are never revealed to any party except the Claimant and Verifiers operated directly by the CSP; however, session (temporary) shared secrets may be provided to independent Verifiers by the CSP. Approved cryptographic techniques are used for all operations. All sensitive data transfers are cryptographically authenticated using keys bound to the authentication process.

At Level 4, "bearer" assertions (as defined in Section 9) are not used to establish the identity of the Claimant to the Relying Party (RP). "Holder-of-key" assertions (as defined in Section 9) may be used, provided that the assertion contains a reference to a key that is possessed by the Subscriber and is cryptographically linked to the Level 4 token used to authenticate to the Verifier. The RP should maintain records of the assertions it receives, to support detection of a compromised verifier impersonating the subscriber.

## Table of Contents

1. Purpose .................................................................................................................... 1
2. Introduction ............................................................................................................. 1
3. Definitions and Abbreviations ................................................................................ 6
4. E-Authentication Model......................................................................................... 16
   4.1. Overview ....................................................................................................... 16
   4.2. Subscribers, Registration Authorities and Credential Service Providers.......... 19
   4.3. Tokens ........................................................................................................... 20
   4.4. Credentials .................................................................................................... 22
   4.5. Authentication Process.................................................................................. 23
   4.6. Assertions ...................................................................................................... 24
   4.7. Relying Parties .............................................................................................. 25
   4.8. Calculating the Overall Authentication Assurance Level.............................. 25
5. Registration and Issuance Processes..................................................................... 27
   5.1. Overview ....................................................................................................... 27
   5.2. Registration and Issuance Threats ................................................................. 28
      5.2.1. Threat Mitigation Strategies ................................................................... 29
   5.3. Registration and Issuance Assurance Levels ................................................ 30
      5.3.1. General Requirements per Assurance Level........................................... 31
      5.3.2. Requirements for Educational and Financial Institutions and Employers 36
      5.3.3. Requirements for PKI Certificates Issued under FPKI and Mapped Policies 38
      5.3.4. Requirements for One-Time Use ............................................................ 38
      5.3.5. Requirements for Derived Credentials ................................................... 39
6. Tokens .................................................................................................................... 40
   6.1. Overview ....................................................................................................... 40
      6.1.1. Single-factor versus Multi-factor Tokens ............................................... 41
      6.1.2. Token Types............................................................................................ 41
      6.1.3. Token Usage ........................................................................................... 43
      6.1.4. Multi-Stage Authentication Using Tokens ............................................. 44
      6.1.5. Assurance Level Escalation .................................................................... 44
   6.2. Token Threats ................................................................................................ 45
      6.2.1. Threat Mitigation Strategies ................................................................... 47
   6.3. Token Assurance Levels ............................................................................... 48
      6.3.1. Requirements per Assurance Level ........................................................ 48
7. Token and Credential Management ...................................................................... 55
   7.1. Overview ....................................................................................................... 55
      7.1.1. Categorizing Credentials......................................................................... 55
      7.1.2. Token and Credential Management Activities ....................................... 56
   7.2. Token and Credential Management Threats ................................................. 58
      7.2.1. Threat Mitigation Strategies ................................................................... 60
   7.3. Token and Credential Management Assurance Levels ................................. 60
      7.3.1. Requirements per Assurance Level ........................................................ 60
      7.3.2. Relationship of PKI Policies to E-Authentication Assurance Levels ....... 66

8. Authentication Process ........................................................................................................ 67
   8.1. Overview ..................................................................................................................... 67
   8.2. Authentication Process Threats .................................................................................. 68
      8.2.1. Other Threats ...................................................................................................... 69
      8.2.2. Threat Mitigation Strategies ............................................................................... 70
      8.2.3. Throttling Mechanisms ....................................................................................... 73
      8.2.4. Phishing and Pharming (Verifier Impersonation): Supplementary Countermeasures ............................................................................................................. 75
   8.3. Authentication Process Assurance Levels .................................................................. 77
      8.3.1. Threat Resistance per Assurance Level .............................................................. 77
      8.3.2. Requirements per Assurance Level .................................................................... 78
9. Assertions .............................................................................................................................. 81
   9.1. Overview ..................................................................................................................... 81
      9.1.1. Cookies ............................................................................................................... 85
      9.1.2. Security Assertion Markup Language (SAML) Assertions ............................... 86
      9.1.3. Kerberos Tickets ................................................................................................. 86
   9.2. Assertion Threats ........................................................................................................ 87
      9.2.1. Threat Mitigation Strategies ............................................................................... 89
   9.3. Assertion Assurance Levels ........................................................................................ 92
      9.3.1. Threat Resistance per Assurance Level .............................................................. 92
      9.3.2. Requirements per Assurance Level .................................................................... 93
10. References ........................................................................................................................... 97
   10.1. General References .................................................................................................... 97
   10.2. NIST Special Publications ......................................................................................... 98
   10.3. Federal Information Processing Standards ................................................................ 99
   10.4. Certificate Policies ..................................................................................................... 99
Appendix A: Estimating Entropy and Strength ......................................................................... 101
   Password Entropy .................................................................................................................. 101
   A.1 Randomly Selected Passwords ................................................................................. 102
   A.2 User Selected Passwords ........................................................................................... 103
   A.3 Other Types of Passwords ......................................................................................... 106
Appendix B: Mapping of Federal PKI Certificate Policies to E-authentication Assurance Levels ......................................................................................................................................... 109

# 1. Purpose

This recommendation provides technical guidelines to agencies for the implementation of electronic authentication (e-authentication).

# 2. Introduction

Electronic authentication (e-authentication) is the process of establishing confidence in user identities electronically presented to an information system. E-authentication presents a technical challenge when this process involves the remote authentication of individual people over a network. This recommendation provides technical guidelines to agencies to allow an individual person to remotely authenticate his/her identity to a Federal Information Technology (IT) system. This recommendation also provides guidelines for Registration Authorities (RAs), Verifiers, Relying Parties (RPs) and Credential Service Providers (CSPs).

Current government systems do not separate the functions of authentication and attribute providers. In some applications, these functions are provided by different parties. While a combined authentication and attribute provider model is used in this document, it does not preclude agencies from separating these functions.

These technical guidelines supplement OMB guidance, *E-Authentication Guidance for Federal Agencies* [OMB M-04-04] and supersede NIST SP 800-63. OMB M-04-04 defines four levels of assurance, Levels 1 to 4, in terms of the consequences of authentication errors and misuse of credentials. Level 1 is the lowest assurance level and Level 4 is the highest. The guidance defines the required level of authentication assurance in terms of the likely consequences of an authentication error. As the consequences of an authentication error become more serious, the required level of assurance increases. The OMB guidance provides agencies with criteria for determining the level of e-authentication assurance required for specific electronic transactions and systems, based on the risks and their likelihood of occurrence.

OMB guidance outlines a 5 step process by which agencies should meet their e-authentication assurance requirements:

1. *Conduct a risk assessment of the government system* – No specific risk assessment methodology is prescribed for this purpose, however the e-RA tool[1] at <http://www.idmanagement.gov/> is an example of a suitable tool and methodology, while NIST Special Publication (SP) 800-30 [SP 800-30] offers a general process for Risk Assessment and Risk Mitigation.

---

[1] At the time of publication, the specific URL for this tool is at <http://www.idmanagement.gov/drilldown.cfm?action=era>. Alternatively, the tool can be found by searching for "Electronic Risk and Requirements Assessment (e-RA)" at <http://www.idmanagement.gov/>.

2. *Map identified risks to the appropriate assurance level* – Section 2.2 of OMB M-04-04 provides the guidance necessary for agencies to perform this mapping.

3. *Select technology based on e-authentication technical guidance* – After the appropriate assurance level has been determined, OMB guidance states that agencies should select technologies that meet the corresponding technical requirements, as specified by this document. Some agencies may possess existing e-authentication technology. Agencies should verify that any existing technology meets the requirements specified in this document.

4. *Validate that the implemented system has met the required assurance level* – As some implementations may create or compound particular risks, agencies should conduct a final validation to confirm that the system achieves the required assurance level for the user-to-agency process. NIST SP 800-53A [SP 800-53A] provides guidelines for the assessment of the implemented system during the validation process. Validation should be performed as part of a security authorization process as described in NIST SP 800-37, Revision 1 [SP 800-37].

5. *Periodically reassess the information system to determine technology refresh requirements* – The agency shall periodically reassess the information system to ensure that the identity authentication requirements continue to be satisfied. NIST SP 800-37, Revision 1 [SP 800-37] provides guidelines on the frequency, depth and breadth of periodic reassessments. As with the initial validation process, agencies should follow the assessment guidelines specified in SP 800-53A [SP 800-53A] for conducting the security assessment.

This document provides guidelines for implementing the third step of the above process. In particular, this document states specific technical requirements for each of the four levels of assurance in the following areas:

- Registration and identity proofing of Applicants (covered in Section 5);
- Tokens (typically a cryptographic key or password) for authentication (covered in Section 6);
- Token and credential management mechanisms used to establish and maintain token and credential information (covered in Section 7);
- Protocols used to support the authentication mechanism between the Claimant and the Verifier (covered in Section 8);
- Assertion mechanisms used to communicate the results of a remote authentication, if these results are sent to other parties (covered in Section 9).

The overall authentication assurance level is determined by the lowest assurance level achieved in any of the areas listed above.

Agencies may adjust the level of assurance using additional risk mitigation measures. Easing credential assurance level requirements may increase the size of the enabled customer pool, but agencies shall ensure that this does not corrupt the system's choice of the appropriate assurance level. Alternatively, agencies may consider partitioning the functionality of an e-authentication enabled application to allow less sensitive functions to be available at a lower level of authentication and attribute assurance, while more sensitive functions are available only at a higher level of assurance.

These technical guidelines cover remote electronic authentication of human users to IT systems over a network. They do not address the authentication of a person who is physically present, for example, for access to buildings, although some credentials and tokens that are used remotely may also be used for local authentication. These technical guidelines establish requirements that Federal IT systems and service providers participating in authentication protocols be authenticated to Subscribers. However, these guidelines do not specifically address machine-to-machine (such as router-to-router) authentication, or establish specific requirements for issuing authentication credentials and tokens to machines and servers when they are used in e-authentication protocols with people.

The paradigm of this document is that individuals are enrolled and undergo a registration process in which their identity is bound to a token. Thereafter, the individuals are remotely authenticated to systems and applications over a network, using the token in an authentication protocol. The authentication protocol allows an individual to demonstrate to a Verifier that he or she has possession and control of the token[2], in a manner that protects the token secret from compromise by different kinds of attacks. Higher authentication assurance levels require use of stronger tokens, better protection of the token and related secrets from attacks, and stronger registration procedures.

This document focuses on tokens that are difficult to forge because they contain some type of secret information that is not available to unauthorized parties and that is preferably not used in unrelated contexts. Certain authentication technologies, particularly biometrics and knowledge based authentication, use information that is private rather than secret. While they are discussed to a limited degree, they are largely avoided because their security is often weak or difficult to quantify[3], especially in the remote situations that are the primary scope of this document.

Knowledge based authentication achieves authentication by testing the personal knowledge of the individual against information obtained from public databases. As this information is considered private but not actually secret, confidence in the identity of an individual can be hard to achieve. In addition, the complexity and interdependencies of knowledge based authentication systems are difficult to quantify. However, knowledge based authentication techniques are included as part of registration in this document. In addition, pre-registered knowledge techniques are accepted as an alternative to passwords at lower levels of assurance.

---

[2] See Section 3 for the definition of "token" as used in this document, which is consistent with the original version of SP 800-63, but there are a variety of definitions used in the area of authentication.
[3] For example, see article by V. Griffith and M. Jakobsson, entitled "Messin' with Texas – Deriving Mother's Maiden Names Using Public Records," in *RSA CryptoBytes*, Winter 2007.

Biometric characteristics do not constitute secrets suitable for use in the conventional remote authentication protocols addressed in this document either. In the local authentication case, where the Claimant is observed by an attendant and uses a capture device controlled by the Verifier, authentication does not require that biometrics be kept secret. This document supports the use of biometrics to "unlock" conventional authentication tokens, to prevent repudiation of registration, and to verify that the same individual participates in all phases of the registration process.

This document identifies minimum technical requirements for remotely authenticating identity. Agencies may determine based on their risk analysis that additional measures are appropriate in certain contexts. In particular, privacy requirements and legal risks may lead agencies to determine that additional authentication measures or other process safeguards are appropriate. When developing e-authentication processes and systems, agencies should consult *OMB Guidance for Implementing the Privacy Provisions of the E-Government Act of 2002* [OMB M-03-22]. See the *Guide to Federal Agencies on Implementing Electronic Processes* [DOJ 2000] for additional information on legal risks, especially those that are related to the need to satisfy legal standards of proof and prevent repudiation, as well as *Use of Electronic Signatures in Federal Organization Transactions* [GSA ESIG].

Additionally, Federal agencies implementing these guidelines should adhere to the requirements of Title III of the E-Government Act, entitled the *Federal Information Security Management Act* [FISMA], and the related NIST standards and guidelines. FISMA directs Federal agencies to develop, document, and implement agency-wide programs to provide information security for the information and information systems that support the operations and assets of the agency. This includes the security authorization of IT systems that support e-authentication. It is recommended that non-Federal entities implementing these guidelines follow equivalent standards of security management, certification and accreditation to ensure the secure operations of their e-authentication systems.

This document has been updated to reflect current (token) technologies and has been restructured to provide a better understanding of the e-authentication architectural model used here. Additional (minimum) technical requirements have been specified for the CSP, protocols utilized to transport authentication information, and assertions if implemented within the e-authentication model. Other changes since NIST SP 800-63 was originally published include:

- Recognition of more types of tokens, including pre-registered knowledge token, look-up secret token, out-of-band token, as well as some terminology changes for more conventional token types;
- Detailed requirements for assertion protocols and Kerberos;
- A new section on token and credential management;
- Simplification of guidelines for password entropy and throttling;
- Emphasis that the document is aimed at Federal IT systems;
- Recognition of different models, including a broader e-authentication model (in contrast to the simpler model common among Federal IT systems shown in Figure 1)

and an additional assertion model, the Proxy Model, presented in Figure 6;
- Clarification of differences between Levels 3 and 4 in Table 12; and
- New guidelines that permit leveraging existing credentials to issue derived credentials.

The subsequent sections present a series of recommendations for the secure implementation of RAs, CSPs, Verifiers, and RPs. It should be noted that secure implementation of any one of these can only provide the desired level of assurance if the others are also implemented securely. Therefore, the following assumptions have been made in this guideline:

- RAs, CSPs, and Verifiers are trusted entities. Agencies implementing any of the above trusted entities have some assurance that all other trusted entities with which the agency interacts are also implemented appropriately for the desired security level.

- The RP is not considered a trusted entity. However, in some authentication systems the Verifier maintains a relationship with the RP to facilitate secure communications and may employ security controls which only attain their full value when the RP acts responsibly. The Subscriber also trusts the RP to properly perform the requested service and to follow all relevant privacy policy.

- It is assumed that there exists a process of certification through which agencies can obtain the above assurance for trusted entities which they do not implement themselves.

- A trusted entity is considered to be implemented appropriately if it complies with the recommendations in this document and does not behave maliciously.

- While it is generally assumed that trusted entities will not behave maliciously, this document does contain some recommendations to reduce and isolate any damage done by a malicious or negligent trusted entity.

## 3. Definitions and Abbreviations

There are a variety of definitions used in the area of authentication. We have kept terms consistent with the original version of SP 800-63. Pay careful attention to how the terms are defined here.

| | |
|---|---|
| Active Attack | An attack on the authentication protocol where the Attacker transmits data to the Claimant, Credential Service Provider, Verifier, or Relying Party. Examples of active attacks include man-in-the-middle, impersonation, and session hijacking. |
| Address of Record | The official location where an individual can be found. The address of record always includes the residential street address of an individual and may also include the mailing address of the individual. In very limited circumstances, an Army Post Office box number, Fleet Post Office box number or the street address of next of kin or of another contact individual can be used when a residential street address for the individual is not available. |
| Approved | Federal Information Processing Standard (FIPS) approved or NIST recommended. An algorithm or technique that is either 1) specified in a FIPS or NIST Recommendation, or 2) adopted in a FIPS or NIST Recommendation. |
| Applicant | A party undergoing the processes of registration and identity proofing. |
| Assertion | A statement from a Verifier to a Relying Party (RP) that contains identity information about a Subscriber. Assertions may also contain verified attributes. |
| Assertion Reference | A data object, created in conjunction with an assertion, which identifies the Verifier and includes a pointer to the full assertion held by the Verifier. |
| Assurance | In the context of OMB M-04-04 and this document, assurance is defined as 1) the degree of confidence in the vetting process used to establish the identity of an individual to whom the credential was issued, and 2) the degree of confidence that the individual who uses the credential is the individual to whom the credential was issued. |
| Asymmetric Keys | Two related keys, a public key and a private key that are used to perform complementary operations, such as encryption and decryption or signature generation and signature verification. |
| Attack | An attempt by an unauthorized individual to fool a Verifier or a Relying Party into believing that the unauthorized individual in question is the Subscriber. |
| Attacker | A party who acts with malicious intent to compromise an information system. |
| Attribute | A claim of a named quality or characteristic inherent in or ascribed to someone or something. (See term in [ICAM] for more information.) |
| Authentication | The process of establishing confidence in the identity of users or information systems. |

| | |
|---|---|
| Authentication Protocol | A defined sequence of messages between a Claimant and a Verifier that demonstrates that the Claimant has possession and control of a valid token to establish his/her identity, and optionally, demonstrates to the Claimant that he or she is communicating with the intended Verifier. |
| Authentication Protocol Run | An exchange of messages between a Claimant and a Verifier that results in authentication (or authentication failure) between the two parties. |
| Authentication Secret | A generic term for any secret value that could be used by an Attacker to impersonate the Subscriber in an authentication protocol.<br><br>These are further divided into *short-term authentication secrets*, which are only useful to an Attacker for a limited period of time, and *long-term authentication secrets*, which allow an Attacker to impersonate the Subscriber until they are manually reset. The token secret is the canonical example of a long term authentication secret, while the token authenticator, if it is different from the token secret, is usually a short term authentication secret. |
| Authenticity | The property that data originated from its purported source. |
| Bearer Assertion | An assertion that does not provide a mechanism for the Subscriber to prove that he or she is the rightful owner of the assertion. The RP has to assume that the assertion was issued to the Subscriber who presents the assertion or the corresponding assertion reference to the RP. |
| Bit | A binary digit: 0 or 1. |
| Biometrics | Automated recognition of individuals based on their behavioral and biological characteristics.<br><br>In this document, biometrics may be used to unlock authentication tokens and prevent repudiation of registration. |
| Certificate Authority (CA) | A trusted entity that issues and revokes public key certificates. |
| Certificate Revocation List (CRL) | A list of revoked public key certificates created and digitally signed by a Certificate Authority. See [RFC 5280]. |
| Challenge-Response Protocol | An authentication protocol where the Verifier sends the Claimant a challenge (usually a random value or a nonce) that the Claimant combines with a secret (such as by hashing the challenge and a shared secret together, or by applying a private key operation to the challenge) to generate a response that is sent to the Verifier. The Verifier can independently verify the response generated by the Claimant (such as by re-computing the hash of the challenge and the shared secret and comparing to the response, or performing a public key operation on the response) and establish that the Claimant possesses and controls the secret. |
| Claimant | A party whose identity is to be verified using an authentication protocol. |

| | |
|---|---|
| Claimed Address | The physical location asserted by an individual (e.g. an applicant) where he/she can be reached. It includes the residential street address of an individual and may also include the mailing address of the individual.<br><br>For example, a person with a foreign passport, living in the U.S., will need to give an address when going through the identity proofing process. This address would not be an "address of record" but a "claimed address." |
| Completely Automated Public Turing test to tell Computers and Humans Apart (CAPTCHA) | An interactive feature added to web-forms to distinguish use of the form by humans as opposed to automated agents. Typically, it requires entering text corresponding to a distorted image or from a sound stream. |
| Cookie | A character string, placed in a web browser's memory, which is available to websites within the same Internet domain as the server that placed them in the web browser.<br><br>Cookies are used for many purposes and may be assertions or may contain pointers to assertions. See Section 9.1.1 for more information. |
| Credential | An object or data structure that authoritatively binds an identity (and optionally, additional attributes) to a token possessed and controlled by a Subscriber.<br><br>While common usage often assumes that the credential is maintained by the Subscriber, this document also uses the term to refer to electronic records maintained by the CSP which establish a binding between the Subscriber's token and identity. |
| Credential Service Provider (CSP) | A trusted entity that issues or registers Subscriber tokens and issues electronic credentials to Subscribers. The CSP may encompass Registration Authorities (RAs) and Verifiers that it operates. A CSP may be an independent third party, or may issue credentials for its own use. |
| Cross Site Request Forgery (CSRF) | An attack in which a Subscriber who is currently authenticated to an RP and connected through a secure session, browses to an Attacker's website which causes the Subscriber to unknowingly invoke unwanted actions at the RP.<br><br>For example, if a bank website is vulnerable to a CSRF attack, it may be possible for a Subscriber to unintentionally authorize a large money transfer, merely by viewing a malicious link in a webmail message while a connection to the bank is open in another browser window. |

| | |
|---|---|
| Cross Site Scripting (XSS) | A vulnerability that allows attackers to inject malicious code into an otherwise benign website. These scripts acquire the permissions of scripts generated by the target website and can therefore compromise the confidentiality and integrity of data transfers between the website and client. Websites are vulnerable if they display user supplied data from requests or forms without sanitizing the data so that it is not executable. |
| Cryptographic Key | A value used to control cryptographic operations, such as decryption, encryption, signature generation or signature verification. For the purposes of this document, key requirements shall meet the minimum requirements stated in Table 2 of NIST SP 800-57 Part 1.<br>See also Asymmetric keys, Symmetric key. |
| Cryptographic Token | A token where the secret is a cryptographic key. |
| Data Integrity | The property that data has not been altered by an unauthorized entity. |
| Derived Credential | A credential issued based on proof of possession and control of a token associated with a previously issued credential, so as not to duplicate the identity proofing process. |
| Digital Signature | An asymmetric key operation where the private key is used to digitally sign data and the public key is used to verify the signature. Digital signatures provide authenticity protection, integrity protection, and non-repudiation. |
| Eavesdropping Attack | An attack in which an Attacker listens passively to the authentication protocol to capture information which can be used in a subsequent active attack to masquerade as the Claimant. |
| Electronic Authentication (E-Authentication) | The process of establishing confidence in user identities electronically presented to an information system. |
| Entropy | A measure of the amount of uncertainty that an Attacker faces to determine the value of a secret. Entropy is usually stated in bits. See Appendix A. |
| Extensible Mark-up Language (XML) | Extensible Markup Language, abbreviated XML, describes a class of data objects called XML documents and partially describes the behavior of computer programs which process them. |
| Federal Bridge Certification Authority (FBCA) | The FBCA is the entity operated by the Federal Public Key Infrastructure (FPKI) Management Authority that is authorized by the Federal PKI Policy Authority to create, sign, and issue public key certificates to Principal CAs. |
| Federal Information Security Management Act (FISMA) | Title III of the E-Government Act requiring each federal agency to develop, document, and implement an agency-wide program to provide information security for the information and information systems that support the operations and assets of the agency, including those provided or managed by another agency, contractor, or other source. |

| | |
|---|---|
| Federal Information Processing Standard (FIPS) | Under the Information Technology Management Reform Act (Public Law 104-106), the Secretary of Commerce approves standards and guidelines that are developed by the National Institute of Standards and Technology (NIST) for Federal computer systems. These standards and guidelines are issued by NIST as Federal Information Processing Standards (FIPS) for use government-wide. NIST develops FIPS when there are compelling Federal government requirements such as for security and interoperability and there are no acceptable industry standards or solutions. See background information for more details. FIPS documents are available online through the FIPS home page: http://www.nist.gov/itl/fips.cfm |
| Guessing Entropy | A measure of the difficulty that an Attacker has to guess the average password used in a system. In this document, entropy is stated in bits. When a password has n-bits of guessing entropy then an Attacker has as much difficulty guessing the average password as in guessing an n-bit random quantity. The Attacker is assumed to know the actual password frequency distribution. See Appendix A. |
| Hash Function | A function that maps a bit string of arbitrary length to a fixed length bit string. Approved hash functions satisfy the following properties:<br>1. (One-way) It is computationally infeasible to find any input that maps to any pre-specified output, and<br>2. (Collision resistant) It is computationally infeasible to find any two distinct inputs that map to the same output. |
| Holder-of-Key Assertion | An assertion that contains a reference to a symmetric key or a public key (corresponding to a private key) held by the Subscriber. The RP may authenticate the Subscriber by verifying that he or she can indeed prove possession and control of the referenced key. |
| Identity | A set of attributes that uniquely describe a person within a given context. |
| Identity Proofing | The process by which a CSP and a Registration Authority (RA) collect and verify information about a person for the purpose of issuing credentials to that person. |
| Kerberos | A widely used authentication protocol developed at MIT. In "classic" Kerberos, users share a secret password with a Key Distribution Center (KDC). The user, Alice, who wishes to communicate with another user, Bob, authenticates to the KDC and is furnished a "ticket" by the KDC to use to authenticate with Bob.<br><br>When Kerberos authentication is based on passwords, the protocol is known to be vulnerable to off-line dictionary attacks by eavesdroppers who capture the initial user-to- KDC exchange. Longer password length and complexity provide some mitigation to this vulnerability, although sufficiently long passwords tend to be cumbersome for users. |

| | |
|---|---|
| Knowledge Based Authentication | Authentication of an individual based on knowledge of information associated with his or her claimed identity in public databases. Knowledge of such information is considered to be private rather than secret, because it may be used in contexts other than authentication to a Verifier, thereby reducing the overall assurance associated with the authentication process. |
| Man-in-the-Middle Attack (MitM) | An attack on the authentication protocol run in which the Attacker positions himself or herself in between the Claimant and Verifier so that he can intercept and alter data traveling between them. |
| Message Authentication Code (MAC) | A cryptographic checksum on data that uses a symmetric key to detect both accidental and intentional modifications of the data. MACs provide authenticity and integrity protection, but not non-repudiation protection. |
| Min-entropy | A measure of the difficulty that an Attacker has to guess the most commonly chosen password used in a system. In this document, entropy is stated in bits. When a password has n-bits of min-entropy then an Attacker requires as many trials to find a user with that password as is needed to guess an n-bit random quantity. The Attacker is assumed to know the most commonly used password(s). See Appendix A. |
| Multi-Factor | A characteristic of an authentication system or a token that uses more than one authentication factor.<br><br>The three types of authentication factors are something you know, something you have, and something you are. |
| Network | An open communications medium, typically the Internet, that is used to transport messages between the Claimant and other parties. Unless otherwise stated, no assumptions are made about the security of the network; it is assumed to be open and subject to active (i.e., impersonation, man-in-the-middle, session hijacking) and passive (i.e., eavesdropping) attack at any point between the parties (e.g., Claimant, Verifier, CSP or RP). |
| Nonce | A value used in security protocols that is never repeated with the same key. For example, nonces used as challenges in challenge-response authentication protocols must not be repeated until authentication keys are changed. Otherwise, there is a possibility of a replay attack. Using a nonce as a challenge is a different requirement than a random challenge, because a nonce is not necessarily unpredictable. |
| Off-line Attack | An attack where the Attacker obtains some data (typically by eavesdropping on an authentication protocol run or by penetrating a system and stealing security files) that he/she is able to analyze in a system of his/her own choosing. |
| Online Attack | An attack against an authentication protocol where the Attacker either assumes the role of a Claimant with a genuine Verifier or actively alters the authentication channel. |

| Online Guessing Attack | An attack in which an Attacker performs repeated logon trials by guessing possible values of the token authenticator. |
|---|---|
| Passive Attack | An attack against an authentication protocol where the Attacker intercepts data traveling along the network between the Claimant and Verifier, but does not alter the data (i.e., eavesdropping). |
| Password | A secret that a Claimant memorizes and uses to authenticate his or her identity. Passwords are typically character strings. |
| Personal Identification Number (PIN) | A password consisting only of decimal digits. |
| Personal Identity Verification (PIV) Card | Defined by [FIPS 201] as a physical artifact (e.g., identity card, smart card) issued to federal employees and contractors that contains stored credentials (e.g., photograph, cryptographic keys, digitized fingerprint representation) so that the claimed identity of the cardholder can be verified against the stored credentials by another person (human readable and verifiable) or an automated process (computer readable and verifiable). |
| Personally Identifiable Information (PII) | Defined by GAO Report 08-536 as "Any information about an individual maintained by an agency, including (1) any information that can be used to distinguish or trace an individual's identity, such as name, social security number, date and place of birth, mother's maiden name, or biometric records; and (2) any other information that is linked or linkable to an individual, such as medical, educational, financial, and employment information." |
| Pharming | An attack in which an Attacker corrupts an infrastructure service such as DNS (Domain Name Service) causing the Subscriber to be misdirected to a forged Verifier/RP, which could cause the Subscriber to reveal sensitive information, download harmful software or contribute to a fraudulent act. |
| Phishing | An attack in which the Subscriber is lured (usually through an email) to interact with a counterfeit Verifier/RP and tricked into revealing information that can be used to masquerade as that Subscriber to the real Verifier/RP. |
| Possession and control of a token | The ability to activate and use the token in an authentication protocol. |
| Practice Statement | A formal statement of the practices followed by the parties to an authentication process (i.e., RA, CSP, or Verifier). It usually describes the policies and practices of the parties and can become legally binding. |
| Private Credentials | Credentials that cannot be disclosed by the CSP because the contents can be used to compromise the token. (For more discussion, see Section 7.1.1.) |
| Private Key | The secret part of an asymmetric key pair that is used to digitally sign or decrypt data. |

| | |
|---|---|
| Protected Session | A session wherein messages between two participants are encrypted and integrity is protected using a set of shared secrets called session keys.<br>A participant is said to be *authenticated* if, during the session, he, she or it proves possession of a long term token in addition to the session keys, and if the other party can verify the identity associated with that token. If both participants are authenticated, the protected session is said to be *mutually authenticated*. |
| Pseudonym | A false name.<br>In this document, all unverified names are assumed to be pseudonyms. |
| Public Credentials | Credentials that describe the binding in a way that does not compromise the token. (For more discussion, see Section 7.1.1.) |
| Public Key | The public part of an asymmetric key pair that is used to verify signatures or encrypt data. |
| Public Key Certificate | A digital document issued and digitally signed by the private key of a Certificate authority that binds the name of a Subscriber to a public key. The certificate indicates that the Subscriber identified in the certificate has sole control and access to the private key. See also [RFC 5280]. |
| Public Key Infrastructure (PKI) | A set of policies, processes, server platforms, software and workstations used for the purpose of administering certificates and public-private key pairs, including the ability to issue, maintain, and revoke public key certificates. |
| Registration | The process through which an Applicant applies to become a Subscriber of a CSP and an RA validates the identity of the Applicant on behalf of the CSP. |
| Registration Authority (RA) | A trusted entity that establishes and vouches for the identity or attributes of a Subscriber to a CSP. The RA may be an integral part of a CSP, or it may be independent of a CSP, but it has a relationship to the CSP(s). |
| Relying Party (RP) | An entity that relies upon the Subscriber's token and credentials or a Verifier's assertion of a Claimant's identity, typically to process a transaction or grant access to information or a system. |
| Remote | (*As in remote authentication or remote transaction*) An information exchange between network-connected devices where the information cannot be reliably protected end-to-end by a single organization's security controls.<br><br>Note: Any information exchange across the Internet is considered remote. |
| Replay Attack | An attack in which the Attacker is able to replay previously captured messages (between a legitimate Claimant and a Verifier) to masquerade as that Claimant to the Verifier or vice versa. |

| | |
|---|---|
| Risk Assessment | The process of identifying the risks to system security and determining the probability of occurrence, the resulting impact, and additional safeguards that would mitigate this impact. Part of Risk Management and synonymous with Risk Analysis. |
| Salt | A non-secret value that is used in a cryptographic process, usually to ensure that the results of computations for one instance cannot be reused by an Attacker. |
| Secondary Authenticator | A temporary secret, issued by the Verifier to a successfully authenticated Subscriber as part of an assertion protocol. This secret is subsequently used, by the Subscriber, to authenticate to the RP.<br><br>Examples of secondary authenticators include bearer assertions, assertion references, and Kerberos session keys. |
| Secure Sockets Layer (SSL) | An authentication and security protocol widely implemented in browsers and web servers. SSL has been superseded by the newer Transport Layer Security (TLS) protocol; TLS 1.0 is effectively SSL version 3.1. |
| Security Assertion Mark-up Language (SAML) | An XML-based security specification developed by the Organization for the Advancement of Structured Information Standards (OASIS) for exchanging authentication (and authorization) information between trusted entities over the Internet. See [SAML]. |
| SAML Authentication Assertion | A SAML assertion that conveys information from a Verifier to an RP about a successful act of authentication that took place between the Verifier and a Subscriber. |
| Session Hijack Attack | An attack in which the Attacker is able to insert himself or herself between a Claimant and a Verifier subsequent to a successful authentication exchange between the latter two parties. The Attacker is able to pose as a Subscriber to the Verifier or vice versa to control session data exchange. Sessions between the Claimant and the Relying Party can also be similarly compromised. |
| Shared Secret | A secret used in authentication that is known to the Claimant and the Verifier. |
| Social Engineering | The act of deceiving an individual into revealing sensitive information by associating with the individual to gain confidence and trust. |
| Special Publication (SP) | A type of publication issued by NIST. Specifically, the Special Publication 800-series reports on the Information Technology Laboratory's research, guidelines, and outreach efforts in computer security, and its collaborative activities with industry, government, and academic organizations. |
| Strongly Bound Credentials | Credentials that describe the binding between a user and token in a tamper-evident fashion. (For more discussion, see Section 7.1.1.) |
| Subscriber | A party who has received a credential or token from a CSP. |
| Symmetric Key | A cryptographic key that is used to perform both the cryptographic operation and its inverse, for example to encrypt and decrypt, or create a message authentication code and to verify the code. |

| Token | Something that the Claimant possesses and controls (typically a cryptographic module or password) that is used to authenticate the Claimant's identity. |
|---|---|
| Token Authenticator | The output value generated by a token. The ability to generate valid token authenticators on demand proves that the Claimant possesses and controls the token. Protocol messages sent to the Verifier are dependent upon the token authenticator, but they may or may not explicitly contain it. |
| Token Secret | The secret value, contained within a token, which is used to derive token authenticators. |
| Transport Layer Security (TLS) | An authentication and security protocol widely implemented in browsers and web servers. TLS is defined by [RFC 2246], [RFC 3546], and [RFC 5246]. TLS is similar to the older Secure Sockets Layer (SSL) protocol, and TLS 1.0 is effectively SSL version 3.1. NIST SP 800-52, *Guidelines for the Selection and Use of Transport Layer Security (TLS) Implementations* specifies how TLS is to be used in government applications. |
| Trust Anchor | A public or symmetric key that is trusted because it is directly built into hardware or software, or securely provisioned via out-of-band means, rather than because it is vouched for by another trusted entity (e.g. in a public key certificate). |
| Unverified Name | A Subscriber name that is not verified as meaningful by identity proofing. |
| Valid | In reference to an ID, the quality of not being expired or revoked. |
| Verified Name | A Subscriber name that has been verified by identity proofing. |
| Verifier | An entity that verifies the Claimant's identity by verifying the Claimant's possession and control of a token using an authentication protocol. To do this, the Verifier may also need to validate credentials that link the token and identity and check their status. |
| Verifier Impersonation Attack | A scenario where the Attacker impersonates the Verifier in an authentication protocol, usually to capture information that can be used to masquerade as a Claimant to the real Verifier. |
| Weakly Bound Credentials | Credentials that describe the binding between a user and token in a manner than can be modified without invalidating the credential. (For more discussion, see Section 7.1.1.) |
| Zeroize | Overwrite a memory location with data consisting entirely of bits with the value zero so that the data is destroyed and not recoverable. This is often contrasted with deletion methods that merely destroy reference to data within a file system rather than the data itself. |
| Zero-knowledge Password Protocol | A password based authentication protocol that allows a claimant to authenticate to a Verifier without revealing the password to the Verifier. Examples of such protocols are EKE, SPEKE and SRP. |

## 4. E-Authentication Model

### 4.1. Overview

In accordance with [OMB M-04-04], e-authentication is the process of establishing confidence in user identities electronically presented to an information system. Systems can use the authenticated identity to determine if that individual is authorized to perform an electronic transaction. In most cases, the authentication and transaction take place across an open network such as the Internet; however, in some cases access to the network may be limited and access control decisions may take this into account.

The e-authentication model used in these guidelines reflects current technologies and architectures used in government. More complex models that separate functions, such as issuing credentials and providing attributes, among larger numbers of parties are also possible and may have advantages in some classes of applications. While a simpler model is used in this document, it does not preclude agencies from separating these functions.

E-authentication begins with *registration*. The usual sequence for registration proceeds as follows. An *Applicant* applies to a *Registration Authority (RA)* to become a *Subscriber* of a *Credential Service Provider (CSP)*. If approved, the Subscriber is issued a *credential* by the CSP which binds a *token* to an identifier (and possibly one or more attributes that the RA has verified). The token may be issued by the CSP, generated directly by the Subscriber, or provided by a third party. The CSP registers the token by creating a *credential* that binds the token to an identifier and possibly other attributes that the RA has verified. The token and credential may be used in subsequent authentication events.

The name specified in a credential may either be a *verified name* or an *unverified name*. If the RA has determined that the name is officially associated with a real person and the Subscriber is the person who is entitled to use that identity, the name is considered a verified name. If the RA has not verified the Subscriber's name, or the name is known to differ from the official name, the name is considered a *pseudonym*. The process used to verify a Subscriber's association with a name is called identity proofing, and is performed by an RA that registers Subscribers with the CSP. At Level 1, identity proofing is not required so names in credentials and assertions are assumed to be pseudonyms. At Level 2, identity proofing is required, but the credential may assert the verified name or a pseudonym. In the case of a pseudonym, the CSP shall retain the name verified during registration. Level 2 credentials and assertions shall specify whether the name is a verified name or a pseudonym. This information assists *Relying Parties (RPs)* in making access control or authorization decisions. In most cases, only verified names may be specified in credentials and assertions at Levels 3 and 4.[4] (The required use of a verified

---

[4] Note that [FIPS 201] permits authorized pseudonyms in limited cases and does not differentiate between credentials using authorized pseudonyms. Nothing in these guidelines should be interpreted as contravening the contents of the FIPS or constraining the use of these authorized pseudonymous credentials. See Appendix B for the level of PIV credentials.

name at higher levels of assurance is derived from OMB M-04-04 and is specific to Federal IT systems, rather than a general e-authentication requirement.)

In this document, the party to be authenticated is called a *Claimant* and the party verifying that identity is called a *Verifier*. When a *Claimant* successfully demonstrates possession and control of a token to a *Verifier* through an *authentication protocol,* the Verifier can verify that the Claimant is the Subscriber named in the corresponding credential. The Verifier passes on an assertion about the identity of the Subscriber to the Relying Party (RP). That assertion includes identity information about a Subscriber, such as the Subscriber name, an identifier assigned at registration, or other Subscriber attributes that were verified in the registration process (subject to the policies of the CSP and the needs of the application). Where the Verifier is also the RP, the assertion may be implicit. The RP can use the authenticated information provided by the Verifier to make access control or authorization decisions.

Authentication establishes confidence in the Claimant's identity, and in some cases in the Claimant's personal attributes (for example the Subscriber is a US Citizen, is a student at a particular university, or is assigned a particular number or code by an agency or organization). Authentication does not determine the Claimant's authorizations or access privileges; this is a separate decision. RPs (e.g., government agencies) will use a Subscriber's authenticated identity and attributes with other factors to make access control or authorization decisions.

As part of authentication, mechanisms such as device identity or geo-location could be used to identify or prevent possible authentication false positives. While these mechanisms do not directly increase the assurance level for authentication, they can enforce security policies and mitigate risks. In many cases, the authentication process and services will be shared by many applications and agencies. However, it is the individual agency or application acting as the RP that shall make the decision to grant access or process a transaction based on the specific application requirements.

The various entities and interactions that comprise the e-authentication model used here are illustrated below in Figure 1. The shaded box on the left shows the registration, credential issuance, maintenance activities, and the interactions between the Subscriber/Claimant, the RA and the CSP. The usual sequence of interactions is as follows:

1. An individual Applicant applies to an RA through a registration process.
2. The RA identity proofs that Applicant.
3. On successful identity proofing, the RA sends the CSP a registration confirmation message.
4. A secret token and a corresponding credential are established between the CSP and the new Subscriber.

5. The CSP maintains the credential, its status, and the registration data collected for the lifetime of the credential (at a minimum).[5] The Subscriber maintains his or her token.

Other sequences are less common, but could also achieve the same functional requirements.

The shaded box on the right side of Figure 1 shows the entities and the interactions related to using a token and credential to perform e-authentication. When the Subscriber needs to authenticate to perform a transaction, he or she becomes a Claimant to a Verifier. The interactions are as follows:

1. The Claimant proves to the Verifier that he or she possesses and controls the token through an authentication protocol.

2. The Verifier interacts with the CSP to validate the credential that binds the Subscriber's identity to his or her token.

3. If the Verifier is separate from the RP (application), the Verifier provides[6] an assertion about the Subscriber to the RP, which uses the information in the assertion to make an access control or authorization decision.

4. An authenticated session is established between the Subscriber and the RP.

In some cases the Verifier does not need to directly communicate with the CSP to complete the authentication activity (e.g., some uses of digital certificates). Therefore, the dashed line between the Verifier and the CSP represents a logical link between the two entities rather than a physical link. In some implementations, the Verifier, RP and the CSP functions may be distributed and separated as shown in Figure 1; however, if these functions reside on the same platform, the interactions between the components are local messages between applications running on the same system rather than protocols over shared untrusted networks.

As noted above, CSPs maintain status information about credentials they issue. CSPs will generally assign a finite lifetime when issuing credentials to limit the maintenance period. When the status changes, or when the credentials near expiration, credentials may be renewed or re-issued; or, the credential may be revoked and/or destroyed. Typically, the Subscriber authenticates to the CSP using his or her existing, unexpired token and credential in order to request re-issuance of a new token and credential. If the Subscriber fails to request token and credential re-issuance prior to their expiration or revocation, he or she may be required to repeat the registration process to obtain a new token and credential. The CSP may choose to accept a request during a grace period after expiration.

---

[5] CSPs may be required to maintain this information beyond the lifetime of the credential to support auditing or satisfy archiving requirements.
[6] Many assertion protocols require assertions to be forwarded through the Claimant's local system before reaching the Relying Party. For Details, see Section 10.

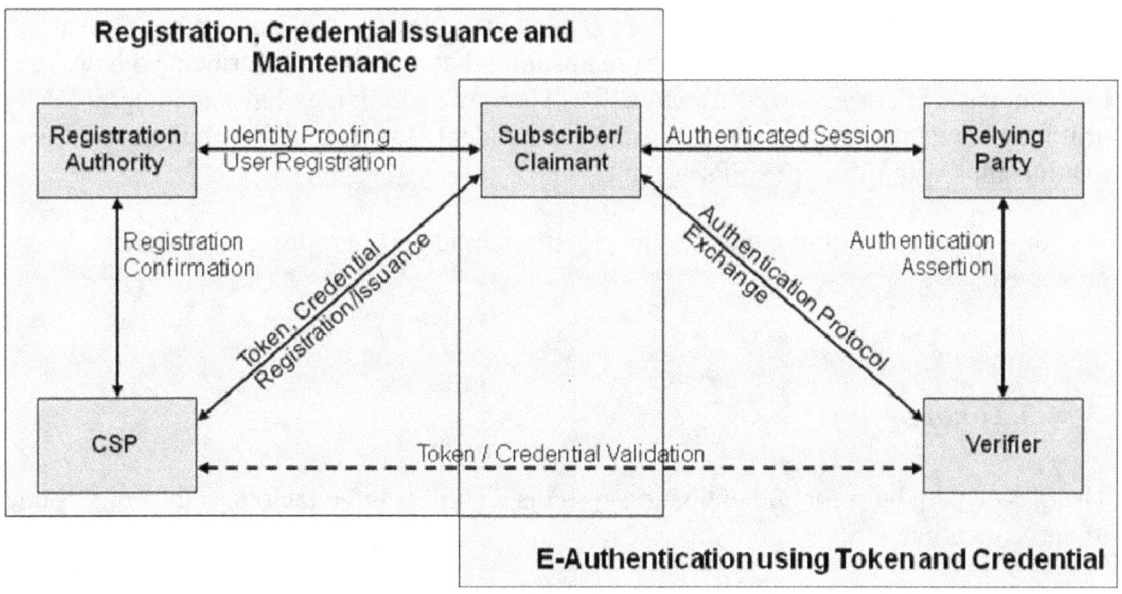

Figure 1 - *The NIST SP 800-63-1 E-Authentication Architectural Model*

### 4.2. Subscribers, Registration Authorities and Credential Service Providers

The previous section introduced the different participants in the conceptual e-authentication model. This section provides additional details regarding the relationships and responsibilities of the participants involved with Registration, Credential Issuance and Maintenance (see the box on the left hand side of Figure 1).

A user may be referred to as the Applicant, Subscriber, or Claimant, depending on the stage in the lifecycle of the credential. An Applicant requests credentials from a CSP. If the Applicant is approved and credentials are issued by a CSP, the user is then termed a Subscriber of that CSP. A user may be a Subscriber of multiple CSPs to obtain appropriate credentials for different applications. A Claimant participates in an authentication protocol with a Verifier to prove they are the Subscriber named in a particular credential.

The CSP establishes a mechanism to uniquely identify each Subscriber, register the Subscriber's tokens, and track the credentials issued to that Subscriber for each token. The Subscriber may be given credentials to go with the token at the time of registration, or credentials may be generated later as needed. Subscribers have a duty to maintain control of their tokens and comply with the responsibilities to the CSP. The CSP (or the RA) maintains registration records for each Subscriber to allow recovery of registration records.

There is always a relationship between the RA and CSP. In the simplest and perhaps the most common case, the RA and CSP are separate functions of the same entity. However, an RA might be part of a company or organization that registers Subscribers with an independent CSP, or several different CSPs. Therefore a CSP may have an integral RA, or it may have relationships with multiple independent RAs, and an RA may have relationships with different CSPs as well.

Section 5 specifies requirements for the registration, identity proofing and issuance processes.

### 4.3. Tokens

The classic paradigm for authentication systems identifies three factors as the cornerstone of authentication:

- *Something you know* (for example, a password)
- *Something you have* (for example, an ID badge or a cryptographic key)
- *Something you are* (for example, a fingerprint or other biometric data)

Multi-factor authentication refers to the use of more than one of the factors listed above. The strength of authentication systems is largely determined by the number of factors incorporated by the system. Implementations that use two factors are considered to be stronger than those that use only one factor; systems that incorporate all three factors are stronger than systems that only incorporate two of the factors. (As discussed in Section 4.1, other types of information, such as location data or device identity, may be used by an RP or Verifier to reject or challenge a claimed identity, but they are not considered authentication factors.)

In e-authentication, the base paradigm is slightly different: the Claimant possesses and controls a token that has been registered with the CSP and is used to prove the bearer's identity. The token contains a secret the Claimant can use to prove that he or she is the Subscriber named in a particular credential.[7] In e-authentication, the Claimant authenticates to a system or application over a network by proving that he or she has possession and control of a token. The token provides an output called a token authenticator. This output is used in the authentication process to prove that the Claimant possesses and controls the token (refer to Section 6.1 for more details), demonstrating that the Claimant is the person to whom the token was issued. Depending on the type of token, this authenticator may or may not be unique for individual authentication operations.

---

[7] The stipulation that every token contains a secret is specific to these E-authentication guidelines. As noted elsewhere authentication techniques where the token does not contain a secret may be applicable to authentication problems in other environments (e.g., physical access).

The secrets contained in tokens are based on either *public key pairs* (asymmetric keys) or *shared secrets*.

A *public key* and a related private key comprise a public key pair. The *private key* is stored on the token and is used by the Claimant to prove possession and control of the token. A Verifier, knowing the Claimant's public key through some credential (typically a *public key certificate*), can use an authentication protocol to verify the Claimant's identity, by proving that the Claimant has possession and control of the associated private key token.

Shared secrets stored on tokens may be either *symmetric keys* or passwords. While they can be used in similar protocols, one important difference between the two is how they relate to the subscriber. While symmetric keys are generally stored in hardware or software that the Subscriber controls, passwords tend to be memorized by the Subscriber. As such, keys are something the Subscriber has, while passwords are something he or she knows. Since passwords are committed to memory, they usually do not have as many possible values as cryptographic keys, and, in many protocols, are vulnerable to network attacks that are impractical for keys. Moreover the entry of passwords into systems (usually through a keyboard) presents the opportunity for very simple keyboard logging attacks, and it may also allow those nearby to learn the password by watching it being entered. Therefore, keys and passwords demonstrate somewhat separate authentication properties (something you have rather than something you know). However, when using either public key pairs or shared secrets, the Subscriber has a duty to maintain exclusive control of his or her token, since possession and control of the token is used to authenticate the Claimant's identity. Token threats are discussed more in Section 6.2.

In this document, e-authentication tokens always contain a secret. So, some of the classic authentication factors do not apply directly to e-authentication. For example, an ID badge is *something you have*, and is useful when authenticating to a human (e.g., a guard), but is not a token for e-authentication. Authentication factors classified as *something you know* are not necessarily secrets, either. Knowledge based authentication, where the claimant is prompted to answer questions that can be confirmed from public databases, also does not constitute an acceptable secret for e-authentication. More generally, *something you are* does not generally constitute a secret. Accordingly, this recommendation does not permit the use of biometrics as a token.

However, this recommendation does accept the notional model that authentication systems that incorporate all three factors offer better security than systems that only incorporate two of the factors. An e-authentication system may incorporate multiple factors in either of two ways. The system may be implemented so that multiple factors are presented to the Verifier, or some factors may be used to protect a secret that will be presented to the Verifier. If multiple factors are presented to the Verifier, each will need to be a token (and therefore contain a secret). If a single factor is presented to the Verifier, the additional factors are used to protect the token and need not themselves be tokens.

For example, consider a piece of hardware (the token) which contains a cryptographic key (the token secret) where access is protected with a fingerprint. When used with the biometric, the cryptographic key produces an output (the token authenticator) which is used in the authentication process to authenticate the Claimant. An impostor must steal the encrypted key (by stealing the hardware) and replicate the fingerprint to use the token. This specification considers such a device to effectively provide two factor authentication, although the actual authentication protocol between the Verifier and the Claimant simply proves possession of the key.

As noted above, biometrics do not constitute acceptable secrets for e-authentication, but they do have their place in this specification. Biometric characteristics are unique personal attributes that can be used to verify the identity of a person who is physically present at the point of verification. They include facial features, fingerprints, DNA, iris and retina scans, voiceprints and many other characteristics. This publication recommends that biometrics be used in the registration process for higher levels of assurance to later help prevent a Subscriber who is registered from repudiating the registration, to help identify those who commit registration fraud, and to unlock tokens.

Section 6 provides guidelines on the various types of tokens that may be used for electronic authentication.

### 4.4. Credentials

As described in the preceding sections, e-authentication credentials bind a token to the Subscriber's name as part of the issuance process. Credentials are issued and maintained by the CSP; Verifiers use the credentials to authenticate the Claimant's identity based on possession and control of the corresponding token. This section provides additional background regarding the relationship of credentials in the e-authentication model with traditional (paper) credentials and describes common e-authentication credentials.

Paper credentials are documents that attest to the identity or other attributes of an individual or entity called the subject of the credentials. Some common paper credentials include passports, birth certificates, driver's licenses, and employee identity cards. The authenticity of paper credentials is established in a variety of ways: traditionally perhaps by a signature or a seal, special papers and inks, high quality engraving, and today by more complex mechanisms, such as holograms, that make the credentials recognizable and difficult to copy or forge. In some cases, simple possession of the credentials is sufficient to establish that the physical holder of the credential is indeed the subject of the credentials. More commonly, the credentials contain information such as the subject's description, a picture of the subject or the handwritten signature of the subject, which can be used to authenticate that the holder of the credentials is indeed the subject of the credentials. When these paper credentials are presented in-person, the information contained in those credentials can be checked to verify that the physical holder of the credential is the subject.

E-authentication credentials may be considered the electronic analog to paper credentials. In both cases, a valid credential authoritatively binds an identity to the necessary information for verifying that a person is entitled to claim that identity. However, the use cases differ in several significant aspects.

The Subject simply possesses and presents the paper credentials in most authentication scenarios. Since they are generally easy to copy, mere possession of a valid electronic credential is rarely a sufficient basis for successful authentication. The e-authentication Claimant possesses a token and presents a token authenticator, but is not necessarily in possession of the electronic credentials. For example, password database entries are considered to be credentials for the purpose of this document but are possessed by the Verifier. X.509 public key certificates are a classic example of credentials the Claimant can (and often does) possess.

As was the case for paper credentials, in order to authenticate a Claimant using an electronic credential, the Verifier shall also validate the credential itself (i.e. confirm that the credential was issued by an authorized CSP and has not subsequently expired or been revoked.) There are two ways this can be done: If the credential has been signed by the CSP, the verifier can validate it by checking the signature. Otherwise, validation may be done interactively by querying the CSP directly through a secure protocol.

In the remainder of this document, the term "credentials" refers to electronic credentials unless explicitly noted. Section 7 provides guidelines for token and credential management activities that are applicable to electronic authentication.

### 4.5. Authentication Process

The authentication process begins with the Claimant demonstrating possession and control of a token that is bound to the asserted identity to the Verifier through an authentication protocol. Once possession and control has been demonstrated, the Verifier verifies that the credential remains valid, usually by interacting with the CSP.

The exact nature of the interaction between the Verifier and the Claimant during the authentication protocol is extremely important in determining the overall security of the system. Well designed protocols can protect the integrity and confidentiality of traffic between the Claimant and the Verifier both during and after the authentication exchange, and it can help limit the damage that can be done by an Attacker masquerading as a legitimate Verifier. Additionally, mechanisms located at the Verifier can mitigate online guessing attacks against lower entropy secrets like passwords and PINs by limiting the rate at which an Attacker can make authentication attempts or otherwise delaying incorrect attempts. (Generally, this is done by keeping track of and limiting the number of unsuccessful attempts, since the premise of an online guessing attack is that most attempts will fail.)

The Verifier is a functional role, but is frequently implemented in combination with the CSP and/or the RP. If the Verifier is a separate entity from the CSP, it is often desirable to ensure that the Verifier does not learn the subscriber's token secret in the process of authentication, or at least to ensure that the Verifier does not have unrestricted access to secrets stored by the CSP.

Section 8 provides guidelines for the various types of protocols used by the Verifier to authenticate the Claimant/Subscriber within the e-authentication model.

### 4.6. Assertions

Upon completion of the authentication process, the Verifier generates an assertion containing the result of the authentication and provides it to the RP. If the Verifier is implemented in combination with the RP, the assertion is implicit. If the Verifier is a separate entity from the RP, the assertion is used to pass information about the Claimant or the authentication process from the Verifier to the RP. Assertions may be communicated directly to the RP, or can be forwarded through the Claimant, which has further implications for system design.

An RP trusts an assertion based on the source, the time of creation, and attributes associated with the Claimant. The Verifier is responsible for providing a mechanism by which the integrity of the assertion can be confirmed. The RP is responsible for authenticating the source (the Verifier) and for confirming the integrity of the assertion. When the Verifier passes the assertion through the Claimant, the Verifier shall protect the integrity of the assertion in such a way that it cannot be modified by the Claimant. However, if the Verifier and the RP communicate directly, a protected session may be used to provide the integrity protection. When sending assertions across an open network, the Verifier is responsible for ensuring that any sensitive Subscriber information contained in the assertion can only be extracted by an RP that it trusts to maintain the information's confidentiality.

Examples of assertions include:

- *Cookies* – Cookies are character strings, placed in memory, which are available to websites within the same Internet domain as the server that placed them in the web browser. Cookies are used for many purposes and may be assertions or may contain pointers to assertions.[8]

---

[8] There are specific requirements that agencies must follow when implementing cookies. See OMB Memorandum M-10-22, OMB Guidance for Online Use of Web Measurement and Customization Technologies, available at: http://www.whitehouse.gov/sites/default/files/omb/assets/memoranda_2010/m10-22.pdf as well as OMB Memorandum M-03-22, OMB Guidance for Implementing the Privacy Provisions of the E-Government Act of 2002, available at: http://www.whitehouse.gov/omb/memoranda/m03-22.html.

- *SAML Assertions* – SAML assertions are specified using a mark-up language intended for describing security assertions. They can be used by a Verifier to make a statement to an RP about the identity of a Claimant. SAML assertions may optionally be digitally signed.
- *Kerberos Tickets* – Kerberos Tickets allow a ticket granting authority to issue session keys to two authenticated parties using symmetric key based encapsulation schemes.

Section 9 provides guidelines for the use of assertions in authentication protocols.

### 4.7. Relying Parties

An RP relies on results of an electronic authentication protocol to establish confidence in the identity or attributes of a Subscriber for the purpose of some transaction. RPs will use a Subscriber's authenticated identity, the overall authentication assurance level, and other factors to make access control or authorization decisions. The Verifier and the RP may be the same entity, or they may be separate entities. If they are separate entities, the RP normally receives an assertion from the Verifier. The RP ensures that the assertion came from a Verifier trusted by the RP. The RP also processes any additional information in the assertion, such as personal attributes or expiration times.

Section 9 provides guidelines for the assertions that may be used by RPs to establish confidence in the identities of Claimants when the RP and the Verifier are not co-located.

### 4.8. Calculating the Overall Authentication Assurance Level

The overall authentication assurance level is based on the low watermark of the assurance levels for each of the components of the architecture. For instance, to achieve an overall assurance level of 3:

- The registration and identity proofing process shall, at a minimum, use Level 3 processes or higher.
- The token (or combination of tokens) used shall have an assurance level of 3 or higher.
- The binding between the identity proofing and the token(s), if proofing is done separately from token issuance, shall be established at level 3.
- The authentication protocols used shall have a Level 3 assurance level or higher.
- The token and credential management processes shall use a Level 3 assurance level or higher.

- Authentication assertions (if used) shall have a Level 3 assurance or higher.

The low watermark is the basis for the overall level because the lowest level will likely be the target of the Attacker. For example, if a system uses a token for authentication that has Level 2 assurance, but uses other mechanisms that have Level 3 assurance, the Attacker will likely focus on gaining access to the token since it is easier to attack a system component meeting assurance Level 2 rather than attacking those meeting assurance Level 3. (See Sections 5 through 9 for information on assurance levels for each area.)

## 5. Registration and Issuance Processes

### 5.1. Overview

In the registration process, an Applicant undergoes identity proofing by a trusted RA. If the RA is able to verify the Applicant's identity, the CSP registers or gives the Applicant a token and issues a credential as needed to bind that token to the identity or some related attribute. The Applicant is now a Subscriber of the CSP and may use the token as a Claimant in an authentication protocol. This section describes the requirements for registration and for token and credential issuance.

The RA can be a part of the CSP, or the RA can be a separate and independent entity; however, a trusted relationship always exists between the RA and CSP. The RA or CSP maintain records of the registration. The RA and CSP can provide services on behalf of an organization or may provide services to the public. The processes and mechanisms available to the RA for identity proofing may differ as a result. Where the RA operates on behalf of an organization, the identity proofing process may be able to leverage a pre-existing relationship (e.g., the Applicant is an employee or student). Where the RA provides services to the public, the identity proofing process is generally limited to confirming publicly available information and previously issued credentials.

The registration and identity proofing processes are designed based on the required assurance level, to ensure that the RA/CSP knows the true identity of the Applicant. Specifically, the requirements include measures to ensure that:

- A person with the Applicant's claimed attributes exists, and those attributes are sufficient to uniquely identify a single person;
- The Applicant whose token is registered is in fact the person who is entitled to the identity;
- It is difficult for the Claimant to later repudiate the registration and dispute an authentication using the Subscriber's token.

An Applicant may appear in person to register, or the Applicant may register remotely. Somewhat different processes and mechanisms apply to identity proofing in each case. Remote registration is limited to Levels 1 through 3.

After successful identity proofing of the Applicant, the RA registers the Applicant, and then the CSP is responsible for token and credential issuance for the new Subscriber (additional CSP responsibilities are discussed further in Section 7). Issuance includes creation of the token. Depending on the type of token being used, the CSP will either create a new token and supply the token to the Subscriber, or require the Subscriber to register a token that the Applicant already possesses or has newly created. In either case, the mechanism for transporting the token from the token origination point to the Subscriber may need to be secured to ensure that the confidentiality and integrity of the newly established token is maintained and that token is in possession of correct Applicant.

The CSP is also responsible for the creation of a credential that binds the Subscriber's identity to his or her token. Optionally, the CSP may include other verified attributes about the Subscriber within the credential, such as his or her organizational affiliation, policies, or constraints for token use.

In models where the registration and identity proofing take place separately from credential issuance, the CSP is responsible for verifying that the credential is being issued to the same person who was identity proofed by the RA. In this model, issuance must be strongly bound to registration and identity proofing so that an Attacker cannot pose as a newly registered Subscriber and attempt to collect a token/credential meant for the actual Subscriber. This attack, and similar attacks, can be thwarted by the methods described in Section 5.3.1 (below Table 3), which describes which techniques are considered appropriate for establishing the necessary binding at the various assurance levels.

### 5.2. Registration and Issuance Threats

There are two general categories of threats to the registration process: impersonation and either compromise or malfeasance of the infrastructure (RAs and CSPs). This recommendation concentrates on addressing impersonation threats. Infrastructure threats are addressed by normal computer security controls (e.g., separation of duties, record keeping, independent audits) and are outside the scope of this document[9].

The threats to the issuance process include impersonation attacks and threats to the transport mechanism for the token and credential issuance. Table 1 lists the threats related to registration and issuance.

---

[9] See NIST SP800-53, *Recommended Security Controls For Federal Information Systems* for appropriate security controls.

Table 1 - Registration and Issuance Threats

| Activity | Threat/Attack | Example |
|---|---|---|
| Registration[10] | Impersonation of claimed identity | An Applicant claims an incorrect identity by using a forged driver's license. |
| | Repudiation of registration | A Subscriber denies registration, claiming that he or she did not register that token. |
| Issuance | Disclosure | A key created by the CSP for a Subscriber is copied by an Attacker as it is transported from the CSP to the Subscriber during token issuance. |
| | Tampering | A new password created by the Subscriber is modified by an Attacker as it is being submitted to the CSP during the credential issuance phase. |
| | Unauthorized issuance | A person claiming to be the Subscriber (but in reality is not the Subscriber) is issued credentials for that Subscriber. |

### 5.2.1. Threat Mitigation Strategies

Registration threats can be deterred by making impersonation more difficult to accomplish or increasing the likelihood of detection. This recommendation deals primarily with methods for making impersonation more difficult; however, it does prescribe certain methods and procedures that may help to prove who carried out an impersonation. At each level, methods are employed to determine that a person with the claimed identity exists, that the Applicant is the person who is entitled to the claimed identity, and that the Applicant cannot later repudiate the registration. As the level of assurance increases, the methods employed provide increasing resistance to casual, systematic and insider impersonation. Table 2 lists strategies for mitigating threats to the registration and issuance processes.

---

[10] Some impostors may attempt to register as any Subscriber in the system and other impostors may wish to register as a specific Subscriber.

Table 2 - Registration and Issuance Threat Mitigation Strategies

| Activity | Threat/Attack | Mitigation Strategy |
|---|---|---|
| Registration | Impersonation of claimed identity | RAs request documentation that provides a specified level of confidence (or assurance) in the identity of the Applicant and makes it more difficult for imposters to successfully pass the identity proofing step. Government issued documents such as driver's licenses, and passports presented by the Applicant are often used to assert the identity of the Applicant. |
| | | Have the Applicant provide non-government issued documentation (e.g. electricity bills in the name of the Applicant with the current address of the Applicant printed on the bill, or a credit card bill) to help in achieving a higher level of confidence in the identity of the Applicant. |
| | Repudiation of registration | Have the Applicant sign a form acknowledging participation in the registration activity. |
| Issuance | Disclosure | Issue the token in person, physically mail it in a sealed envelope to a secure location, or use a protected session to send the token electronically. |
| | Tampering | Issue credentials in person, physically mailing storage media in a sealed envelope, or through the use of a communication protocol that protects the integrity of the session data. |
| | | Establish a procedure that allows the Subscriber to authenticate the CSP as the source of any token and credential data that he or she may receive. |
| | Unauthorized issuance | Establish procedures to ensure that the individual who receives the token is the same individual who participated in the registration procedure. |
| | | Implement a dual-control issuance process that ensures two independent individuals shall cooperate in order to issue a token and/or credential. |

## 5.3. Registration and Issuance Assurance Levels

The following sections list the NIST recommendations for registration and issuance for the four levels corresponding to the OMB guidance. As noted in the OMB guidance, Levels 1 and 2 recognize the use of pseudonymous credentials. When pseudonymous credentials are used to imply membership in a group, the level of proofing shall be consistent with the requirements for the credential of that level. Explicit requirements for registration processes for pseudonymous credentials are not specified, as they are unique to the membership criteria for each specific group.

### 5.3.1. General Requirements per Assurance Level

For levels 2 and above records of registration shall be maintained either by the RA or by the CSP, depending on the context. Either the RA or the CSP shall maintain a record of each individual whose identity has been verified and the steps taken to verify his or her identity, including any information collected from the applicant in compliance with the sections below. The CSP shall have the capability to provide records of identity proofing to RPs if required[11]. The identity proofing and registration processes shall be performed according to applicable written policy or *practice statement* that specifies the particular steps taken to verify identities.

For Levels 2 and above, if the RA and CSP are remotely located and communicate over a network, the entire registration transaction between the RA and CSP shall occur over a mutually authenticated protected session. Equivalently, the transaction may consist of time-stamped or sequenced messages signed by their source and encrypted for their recipient. In either case, Approved cryptography is required.

The CSP shall be able to uniquely identify each Subscriber and the associated tokens and the credentials issued to that Subscriber. The CSP shall be capable of conveying this information to Verifiers. At Level 1, the name associated with the Subscriber is provided by the Applicant and accepted without verification. At Level 2, the identifier associated with the Subscriber may be pseudonymous but the RA or CSP shall retain the actual identity of the Subscriber. In addition, pseudonymous Level 2 credentials shall be distinguishable from Level 2 credentials that contain verified names.

At Level 3 and above, the name associated with the Subscriber shall be verified. At all levels, personally identifiable information (PII) collected as part of the registration process shall be protected, and all privacy requirements shall be satisfied.

The following text establishes registration requirements specific to each level, except as noted in the following subsections. There are no level-specific requirements at Level 1. Both in-person and remote registration are permitted for Levels 2 and 3. Explicit requirements are specified for each scenario in Levels 2 and 3. At Level 4, only in-person registration is permitted.

At Level 2 and higher, the Applicant supplies his or her full legal name, an address of record, and date of birth, and may, subject to the policy of the RA or CSP, also supply other PII. Detailed level-by-level identity proofing requirements are stated in Table 3.

---

[11] It is beyond the scope of this document to specify what circumstances make it is necessary and/or appropriate for the CSP to provide this information. Refer to applicable privacy laws, rules of evidence etc.

In some contexts, once an agency has met the minimum registration requirements for an assurance level, the agency may choose to use additional knowledge based authentication methods to increase confidence in the registration process. For example, an Applicant could be asked to supply non-public information on his or her past dealing with the agency that could help confirm the Applicant's identity.

The sensitive data collected during the registration and identity proofing stage shall be protected at all times (i.e., transmission, storage) to ensure their security and confidentiality. Additionally, the results of the identity proofing step (which may include background investigations of the Applicant) have to be protected to ensure source authentication, confidentiality, and integrity.

Table 3 - Identity Proofing Requirements by Assurance Level

| | In-Person | Remote |
|---|---|---|
| **Level 2**[12] | | |
| Basis for issuing credentials | Possession of a valid current primary government picture ID that contains Applicant's picture, and either address of record or nationality of record (e.g., driver's license or Passport) | Possession of a valid current government ID[13] (e.g., a driver's license or Passport) number and a financial or utility account number (e.g. checking account, savings account, utility account, loan or credit card, or tax ID) confirmed via records of either the government ID or account number. Note that confirmation of the financial or utility account may require supplemental information from the applicant. |
| RA and CSP actions | **RA** inspects photo-ID; compares picture to Applicant; and records the ID number, address and date of birth (DoB). (**RA** optionally reviews personal information in records to support issuance process "a" below.)<br><br>If the photo-ID appears valid and the photo matches Applicant then:<br>  a) If personal information in records includes a telephone number or e-mail address, the **CSP** issues credentials in a manner that confirms the ability of the Applicant to receive telephone communications or text message at phone number or e-mail address associated with the Applicant in records. Any secret sent over an unprotected session shall be reset upon first use; or<br>  b) If ID confirms address of record, **RA** authorizes or **CSP** issues credentials. Notice is sent to address of record, or;<br>  c) If ID does not confirm address of record, **CSP** issues credentials in a manner that confirms the claimed address. | • **RA** inspects both ID number and account number supplied by Applicant (e.g., for correct number of digits). Verifies information provided by Applicant including ID number OR account number through record checks either with the applicable agency or institution or through credit bureaus or similar databases, and confirms that: name, DoB, address and other personal information in records are on balance consistent with the application and sufficient to identify a unique individual. For utility account numbers, confirmation shall be performed by verifying knowledge of recent account activity. (This technique may also be applied to some financial accounts.)<br>• Address/phone number confirmation and notification:[14]<br>  a) **CSP** issues credentials in a manner that confirms the ability of the applicant to receive mail at a physical address associated with the Applicant in records; or<br>  b) If personal information in records includes a telephone number or e-mail address, the **CSP** issues credentials in a manner that confirms the ability of the Applicant to receive telephone communications or text message at phone number or e-mail address associated with the Applicant in records. Any secret sent over an unprotected session shall be reset upon first use; or<br>  c) **CSP** issues credentials. **RA** or **CSP** sends notice to an address of record confirmed in the records check.[15] |

---

[12] A token at this Level may also be obtained by authenticating to the CSP using mechanisms at the same or a higher Level (e.g. PIV). See 5.3.5 for more information.

[13] Agencies issuing credentials to foreign nationals residing in foreign countries determine what constitutes a valid Government issued ID as required.

[14] Requirements that use USPS mail for address confirmation and/or notification have a legal basis: Title 18 U.S. Code: Criminal Procedure, Section 1708: Theft or receipt of stolen mail matter generally.

[15] Agencies are encouraged to use methods a) and b) where possible to achieve better security. Method c) is especially weak when not used in combination with knowledge of account activity

| Level 3[12] | | |
|---|---|---|
| Basis for issuing credentials | Possession of verified current primary Government Picture ID that contains Applicant's picture and either address of record or nationality of record (e.g., driver's license or passport) | Possession of a valid Government ID (e.g., a driver's license or Passport) number and a financial or utility account number (e.g., checking account, savings account, utility account, loan or credit card) confirmed via records of both numbers. Note that confirmation of the financial or utility account may require supplemental information from the applicant. |
| RA and CSP actions | *RA* inspects photo-ID and verifies via the issuing government agency or through credit bureaus or similar databases. Confirms that: name, DoB, address and other personal information in record are consistent with the application. Compares picture to Applicant and records ID number.<br><br>If ID is valid and photo matches Applicant, then:<br>a) If personal information in records includes a telephone number, the **CSP** issues credentials in a manner that confirms the ability of the Applicant to receive telephone communications at a number associated with the Applicant in records, while recording the Applicant's voice or using alternative means that establish an equivalent level of non-repudiation; or<br>b) If ID confirms address of record, **RA** authorizes or **CSP** issues credentials. Notice is sent to address of record, or;<br>c) If ID does not confirm address of record, **CSP** issues credentials in a manner that confirms the claimed address. | • **RA** verifies information provided by Applicant including ID number AND account number through record checks either with the applicable agency or institution or through credit bureaus or similar databases, and confirms that: name, DoB, address and other personal information in records are consistent with the application and sufficient to identify a unique individual. For utility account numbers, confirmation shall be performed by verifying knowledge of recent account activity. (This technique may also be applied to some financial accounts.)<br>• Address confirmation:<br>a) **CSP** issues credentials in a manner that confirms the ability of the applicant to receive mail at a physical address associated with the Applicant in records; or[14]<br>b) If personal information in records includes a telephone number, the **CSP** issues credentials in a manner that confirms the ability of the Applicant to receive telephone communications at a number associated with the Applicant in records. **CSP** records the Applicant's voice or using alternative means that establish an equivalent level of non-repudiation. |

| Level 4[12] | | |
|---|---|---|
| Basis for issuing credentials | In-person appearance and verification of:<br>a) a current primary Government Picture ID that contains Applicant's picture, and either address of record or nationality of record (e.g., driver's license or passport), and;<br>b) either a second, independent Government ID document that contains current corroborating information (e.g., either address of record or nationality of record), OR verification of a financial account number (e.g., checking account, savings account, loan or credit card) confirmed via records. | Not Applicable |
| RA and CSP actions | • *Primary Photo ID:*<br>**RA** inspects photo-ID and verifies via the issuing government agency or through credit bureaus or similar databases. Confirms that: name, DoB, address, and other personal information in record are consistent with the application. Compares picture to Applicant and records ID number.<br>• *Secondary Government ID or financial account*<br>a) **RA** inspects secondary Government ID and if apparently valid, confirms that the identifying information is consistent with the primary Photo-ID, or;<br>b) **RA** verifies financial account number supplied by Applicant through record checks or through credit bureaus or similar databases, and confirms that: name, DoB, address, and other personal information in records are on balance consistent with the application and sufficient to identify a unique individual.<br><br>**[Note: Address of record shall be confirmed through validation of either the primary or secondary ID.]**<br><br>• *Current Biometric*<br>**RA** records a current biometric (e.g., photograph or fingerprints) to ensure that Applicant cannot repudiate application.<br>• *Credential Issuance*<br>**CSP** issues credentials in a manner that confirms address of record. | Not Applicable |

Registration, identity proofing, token creation/issuance, and credential issuance are separate processes that can be broken up into a number of separate physical encounters or electronic transactions. (Two electronic transactions are considered to be separate if they are not part of the same protected session.) In these cases, the following methods shall be used to ensure that the same party acts as Applicant throughout the processes:

- At Level 1, there is no specific requirement, however some effort should be made to uniquely identify and track applications.

- At Level 2: For electronic transactions, the Applicant shall identify himself/herself in any new transaction (beyond the first transaction or encounter) by presenting a temporary secret which was established during a prior transaction or encounter, or sent to the Applicant's phone number, email address, or physical address of record. For physical transactions, the Applicant shall identify himself/herself in person by either using a secret as described above, or through the use of a biometric characteristic that was recorded during a prior encounter.

- At Level 3: For electronic transactions, the Applicant shall identify himself/herself in each new electronic transaction by presenting a temporary secret which was established during a prior transaction or encounter, or sent to the Applicant's physical address of record. Permanent secrets shall only be issued to the applicant within a protected session.

  For physical transactions, the Applicant shall identify himself/herself in person by either using a secret as described above, or through the use of a biometric that was recorded during a prior encounter. Temporary secrets shall not be reused. If the CSP issues permanent secrets during a physical transaction, then they shall be loaded locally onto a physical device that is issued in person to the applicant or delivered in a manner that confirms the address of record.

- At Level 4: Only physical transactions apply. The Applicant shall identify himself/herself in person in each new physical transaction through the use of a biometric that was recorded during a prior encounter.[16] If the CSP issues permanent secrets, then they shall be loaded locally onto a physical device that is issued in person or delivered in a manner that confirms the address of record.

A common reason for breaking up the registration process as described above is to allow the subscriber to register or obtain tokens for use in two or more environments. This is permissible as long as the tokens individually meet the appropriate assurance level. However, if the exact number of tokens to be issued is not agreed upon early in the registration process, then the tokens should be distinguishable so that Verifiers will be able to detect whether any suspicious activity occurs during the first few uses of a newly issued token.

If a valid credential has already been issued, the CSP may issue another credential of equivalent or lower assurance. In this case, proof of possession and control of the original token may be substituted for repeating the identity proofing steps. (This is a special case of a derived credential. See Section 5.3.5 for procedures when the derived credential is issued by a different CSP.) Any requirements for credential delivery at the appropriate Level shall still be satisfied.

### 5.3.2. Requirements for Educational and Financial Institutions and Employers

At Level 2, employers and educational institutions who verify the identity of their employees or students by means comparable to those stated above for Level 2 may elect to become an RA or CSP and issue credentials to employees or students, either in-person

---

[16] Special arrangements can be made for Applicants who are unable to provide the required biometrics.

by inspection of a corporate or school issued picture ID, or through online processes, where notification is via the distribution channels normally used for sensitive, personal communications.

Federal law, including the Bank Secrecy Act and the USA PATRIOT Act, impose a duty on financial institutions to "know their customers" and report suspicious transactions to help prevent money laundering and terrorist financing. Many financial institutions are regulated by Federal agencies such as the Office of the Comptroller of the Currency (OCC) or other members of the Federal Financial Institutions Examination Council (FFIEC) and the Securities and Exchanges Commission (SEC). These regulators normally require the institutions to implement a Customer Identification Program.

The following provisions apply to Federally regulated financial institutions, brokerages and dealers subject to such Federal regulation, that implement such a Customer Identification Program:

- At Level 2, such institutions may issue credentials to their customers via the mechanisms normally used for online banking or brokerage credentials and may use online banking or brokerage credentials and tokens as Level 2 e-authentication credentials and tokens, provided they meet the provisions of Sections 6 through 9 for Level 2.

- At Level 3, such institutions may issue credentials to their customers via the mechanisms normally used for online banking or brokerage credentials and may use online banking or brokerage credentials and tokens as Level 3 e-authentication credentials and tokens, provided:

  1. The customers have been in good standing with the institution for a period of at least 1 year prior to the issuance of e-authentication credentials, and

  2. The credentials and tokens meet the provisions of Sections 6 through 9 for Level 3.

- At Level 4, such institutions may issue credentials to their customers via the mechanisms normally used for online banking or brokerage credentials and may use online banking or brokerage credentials and tokens as Level 4 e-authentication credentials, provided:

  1. The customers have appeared in-person before a representative of the financial institution, and the representative has inspected a Government issued primary Photo-ID and compared the picture to the customer; and

  2. The credentials and tokens meet all additional provisions of Section 5, as well as all provisions in Sections 6 through 9 for Level 4, as appropriate.

### 5.3.3. Requirements for PKI Certificates Issued under FPKI and Mapped Policies

The identity proofing and certificate issuance processes specified in the Federal PKI Certificate Policies [FBCA1, FBCA2, FBCA3] are considered equivalent to the requirements specified in Section 5.3.1 in accordance with Appendix B.

At Level 2, agencies may rely on any CA whose policy satisfies the identity proofing and registration requirements specified for Level 2, in addition to any CA cross-certified with the Federal Bridge CA under one of the certificate policies identified in Appendix B as a Level 2 certificate or a policy mapped to one of those policies through cross-certificates. For Levels 3 and 4, agencies shall only accept PKI certificates issued by a CA cross-certified with the Federal Bridge CA under one of the certificate policies identified in Appendix B as a Level 3 or Level 4 certificate or a policy mapped to one of those policies through cross-certificates.

The identity proofing and certificate issuance processes specified in Federal Information Processing Standard (FIPS) 201, 'Personal Identity Verification' [FIPS201], meet and exceed the Level 4 requirements specified in the preceding section.

### 5.3.4. Requirements for One-Time Use

For infrequently used applications, issuance and maintenance of credentials would be prohibitively expensive. Claimants can be authenticated for immediate one-time access to an application for Levels 1 thru 3. At Level 1, there is no requirement for identity proofing before one-time use. At Levels 2 and 3, application owners act as the RA/CSP in the remote registration processes described in Section 5.3.1, using processes that do not require confirmation of the address of record and omitting credential issuance.

For immediate one-time access at Level 2, application owners can use the registration processes specified in Table 3 that:
- Confirm "the ability of the Applicant to receive telephone communications or text message at phone number or e-mail address associated with the Applicant in records"; or
- Subsequently send a "notice to an address of record confirmed in the records check."

For immediate one-time access at Level 3, application owners can use the registration process specified in Table 3 that:
- Confirms "the ability of the Applicant to receive telephone communications at a phone number associated with the Applicant in records while recording the Applicant's voice or using alternative means that establish an equivalent level of non-repudiation."

### 5.3.5. Requirements for Derived Credentials

Where the Applicant already possesses recognized credentials at or above the desired Level of Assurance, the CSP may choose to accept the verified identity of the Applicant as a substitute for identity proofing.

For Level 2 and Level 3, a long term derived credential may be issued based on the validation of an original credential of a higher Assurance Level. At all Levels, the CSP shall verify original credential status and shall verify that the token is possessed and controlled by the Claimant. The status of the original credential should be re-checked at a later date (e.g. after a week) to confirm that it was not compromised at the time of issuance of the derived credential. (This guards against the case where an Attacker requests the desired credential before revocation information can be updated.) Further, the CSP shall record the details of the original credential used as the basis for derived credential issuance. If the derived credential is revoked, the CSP that issued the derived credential may notify the issuer of the original credential, if the reason for revocation might motivate action by the issuer of the original credential and such notification is permitted by applicable law, regulation, and agreements.

At Level 4, a long-term derived credential may be issued based on in person presentation of another Level 4 credential: in addition to the requirements for Levels 2 and Level 3, the biometric data shall be validated against the Claimant, and the binding of the biometric and the original credential shall be confirmed. The CSP shall retain the biometric sample used to validate the Claimant for future reference.

In some cases, there may be a desire to tightly couple the revocation status of the derived credential to the original. In this case, it is the responsibility of the CSP that issued the derived credential to ensure that a tight coupling is maintained. For example, the issuer of the derived credential could regularly monitor the status of the primary credential.[17] [18]

---

[17] This document does not require or prevent CSPs from linking the expiration of the original and derived credentials. However, where the revocation status is tightly coupled, this may simplify revocation procedures.

[18] Requirements for derived credentials issued by the same CSP are at the end of Section 5.3.1.

# 6. Tokens

The concept of a token was introduced in Section 4. This section provides a more in-depth treatment of e-authentication tokens. Section 6.1 describes classes of tokens recognized by this recommendation and how they can be combined in practice. Section 6.2 identifies threats and mitigation strategies applicable to tokens. Section 6.3 maps recognized classes of tokens to assurance levels and identifies any required threat mitigation strategies.

## 6.1. Overview

In the e-authentication context, a token contains a secret to be used in authentication processes. Tokens are possessed by a Claimant and controlled through one or more of the traditional authentication factors (*something you know*, *have*, or *are*). Figure 2 depicts an abstract model for a token.

The outer box shown in Figure 2 is the token. Tokens may exist in hardware (e.g., a smart card), software (e.g., a software cryptographic module), or may only exist in human memory. The inner box represents the token secret that is stored within the token. The output of a token is the *token authenticator*, which is the value that is provided to the protocol stack for transmission to the Verifier to prove that the Claimant possesses and controls the token. The token authenticator may be the token secret, or a transformation of the token secret.

There are two optional inputs to the token: *token input data*; and *token activation data*. Token input data, such as a challenge or nonce, may be required to generate the token authenticator. Token input data may be supplied by the user or be a feature of the token itself (e.g. the clock in an OTP device). Token activation data, such as a PIN or biometric, may be required to activate the token and permit generation of an authenticator. Token activation data is needed when a Claimant controls the token through *something you know* or *something you are*. (Where the token is something you know, such as a password or memorized secret, token activation is implicit.)

The *authenticator* is generated through the use of the token. In the general case, an authenticator is generated by performing a mathematical function using the token secret and one or more optional token input values (a nonce or challenge):

*Authenticator = Function (<token secret> [, <nonce>] [, <challenge>] )*

As noted above, in the trivial case, the authenticator may be the token secret itself (e.g., where the token is a password).

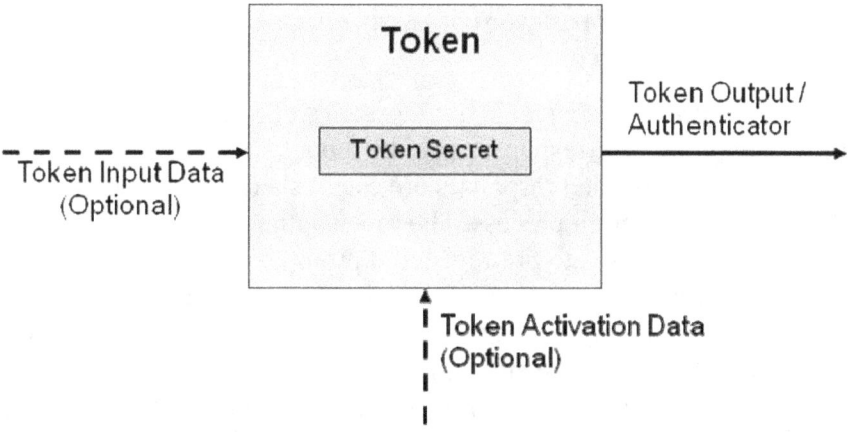

Figure 2 - *Token Model*

### 6.1.1. Single-factor versus Multi-factor Tokens

Tokens are characterized by the number and types of authentication factors that they use. (See Section 4.3 for discussion on three types of authentication factors.) For example, a password is something you know, a biometric is something you are, and a cryptographic identification device is something you have. Tokens may be single-factor or multi-factor tokens as described below:

- *Single-factor Token* – A token that uses one of the three factors to achieve authentication. For example, a password is *something you know*. There are no additional factors required to activate the token, so this is considered single factor.

- *Multi-factor Token* – A token that uses two or more factors to achieve authentication. For example, a private key on a smart card that is activated via PIN is a multi-factor token. The smart card is *something you have*, and *something you know* (the PIN) is required to activate the token.

This document does not differentiate between tokens that require two factors and three factors, as two factors are sufficient to achieve the highest level recognized in this document. Other applications or environments may require such a differentiation.

### 6.1.2. Token Types

These guidelines recognize the following types of tokens for e-authentication.

- *Memorized Secret Token* – A secret shared between the Subscriber and the CSP. Memorized Secret Tokens are typically character strings (e.g., passwords and passphrases) or numerical strings (e.g., PINs.) The token authenticator presented to the Verifier in an authentication process is the secret itself (e.g. the password or passphrase itself). Memorized secret tokens are *something you know*.

- *Pre-registered Knowledge Token* – A series of responses to a set of prompts or challenges. These responses may be thought of as a set of shared secrets. The set of prompts and responses are established by the Subscriber and CSP during the registration process. The token authenticator is the set of memorized responses to pre-registered prompts during a single run of the authentication process. An example of a Pre-registered Knowledge Token would be establishing responses for prompts such as "What was your first pet's name?" During the authentication process, the Claimant is asked to provide the appropriate responses to a subset of the prompts. Alternatively, a Subscriber might select and memorize an image during the registration process. In an authentication process, the Claimant is prompted to identify the correct images from a set(s) of similar images. Transactions from previously authenticated sessions could be accepted as Pre-registered Knowledge Tokens. Pre-registered Knowledge Tokens are *something you know*.

- *Look-up Secret Token* – A physical or electronic token that stores a set of secrets shared between the Claimant and the CSP. The Claimant uses the token to look up the appropriate secret(s) needed to respond to a prompt from the Verifier (the token input). For example, a Claimant may be asked by the Verifier to provide a specific subset of the numeric or character strings printed on a card in table format. The token authenticator is the secret(s) identified by the prompt. Look-up secret tokens are *something you have*.

- *Out of Band Token* – A physical token that is uniquely addressable and can receive a Verifier-selected secret for one-time use. The device is possessed and controlled by the Claimant and supports private communication[19] over a channel that is separate from the primary channel for e-authentication. The token authenticator is the received secret and is presented to the Verifier using the primary channel for e-authentication. For example, a Claimant attempts to log into a website and receives a text message on his or her cellular phone, PDA, pager, or land line (pre-registered with the CSP during the registration phase) with a random authenticator to be presented as a part of the electronic authentication protocol. Out of Band Tokens are *something you have*.

- *Single-factor (SF) One-Time Password (OTP) Device* – A hardware device that supports the spontaneous generation of one-time passwords. This device has an embedded secret that is used as the seed for generation of one-time passwords and does not require activation through a second factor. Authentication is accomplished by providing an acceptable one-time password and thereby proving possession and control of the device. The token authenticator is the one-time password. For example, a one-time password device may display 6 characters at a time. SF OTP devices are *something you have*.

---

[19] Private communication means the Verifier's message is sent directly to the Claimant's device.

- *Single-factor (SF) Cryptographic Device* – a hardware device that performs cryptographic operations on input provided to the device. This device does not require activation through a second factor of authentication. This device uses embedded symmetric or asymmetric cryptographic keys. Authentication is accomplished by proving possession of the device. The token authenticator is highly dependent on the specific cryptographic device and protocol, but it is generally some type of signed message. For example, in TLS, there is a "certificate verify" message. SF Cryptographic Devices are *something you have*.

- *Multi-factor (MF) Software Cryptographic Token* – A cryptographic key is stored on disk or some other "soft" media and requires activation through a second factor of authentication. Authentication is accomplished by proving possession and control of the key. The token authenticator is highly dependent on the specific cryptographic protocol, but it is generally some type of signed message. For example, in TLS, there is a "certificate verify" message. The MF software cryptographic token is *something you have*, and it may be activated by either *something you know* or *something you are*.

- *Multi-factor (MF) One-Time Password (OTP) Device* – A hardware device that generates one-time passwords for use in authentication and which requires activation through a second factor of authentication. The second factor of authentication may be achieved through some kind of integral entry pad, an integral biometric (e.g., fingerprint) reader or a direct computer interface (e.g., USB port). The one-time password is typically displayed on the device and manually input to the Verifier as a password, although direct electronic input from the device to a computer is also allowed. The token authenticator is the one-time password. For example, a one-time password device may display 6 characters at a time. The MF OTP device is *something you have*, and it may be activated by either *something you know* or *something you are*.

- *Multi-factor (MF) Cryptographic Device* – A hardware device that contains a protected cryptographic key that requires activation through a second authentication factor. Authentication is accomplished by proving possession of the device and control of the key. The token authenticator is highly dependent on the specific cryptographic device and protocol, but it is generally some type of signed message. For example, in TLS, there is a "certificate verify" message. The MF Cryptographic device is *something you have*, and it may be activated by either *something you know* or *something you are*.

### 6.1.3. Token Usage

An authentication process may involve a single token, or a combination of two or more tokens, as described below.

- *Single-token authentication* – The Claimant presents a single token authenticator to prove his or her identity to the Verifier. For example, when a Claimant attempts to log into a password protected website, the Claimant enters a username and password. In this instance, only the password would be considered to be a token.

- *Multi-token authentication* – The Claimant presents token authenticators generated by two or more tokens to prove his or her identity to the Verifier. The combination of tokens is characterized by the combination of factors used by the tokens (both inherent in the manifestation of the tokens, and those used to activate the tokens). A Verifier that requires a Claimant to enter a password and use a single-factor cryptographic device is an example of multi-token authentication. The combination is considered multi-factor, since the password is *something you know* and the cryptographic device is *something you have*.

### 6.1.4. Multi-Stage Authentication Using Tokens

Multi-stage authentication processes, which use a single-factor token to obtain a second token, do not constitute multi-factor authentication. The level of assurance associated with the compound solution is the assurance level of the weakest token.

For example, some cryptographic mobility solutions allow full or partial cryptographic keys to be stored on an online server and downloaded to the Claimant's local system after successful authentication using a password or passphrase. Subsequently, the Claimant can use the downloaded software cryptographic token to authenticate to a remote Verifier for e-authentication. This type of solution is considered only as strong as the password provided by the Claimant to obtain the cryptographic token.

### 6.1.5. Assurance Level Escalation

In certain circumstances, it may be desirable to raise the assurance level of an e-authentication session between a Subscriber and an RP in the middle of the application session. This guideline recognizes a special case of multi-token authentication, where a primary token is used to establish a secure session, and a secondary token is used later in the session to present a second token authenticator. Even though the two tokens were not used at the same time, this document recognizes the result as a multi-token authentication scheme (which may upgrade the overall level of assurance). In these authentication scenarios, the level of assurance achieved by the two stages in combination is the same as a multi-token authentication scheme using the same set of tokens. Table 7 describes the highest level of assurance achievable through a combination of two token types.

## 6.2. Token Threats

An Attacker who can gain control of a token will be able to masquerade as the token's owner. Threats to tokens can be categorized based on attacks on the types of authentication factors that comprise the token:

- *Something you have* may be lost, damaged, stolen from the owner or cloned by the Attacker. For example, an Attacker who gains access to the owner's computer might copy a software token. A hardware token might be stolen, tampered with, or duplicated.

- *Something you know* may be disclosed to an Attacker. The Attacker might guess a password or PIN. Where the token is a shared secret, the Attacker could gain access to the CSP or Verifier and obtain the secret value. An Attacker may install malicious software (e.g., a keyboard logger) to capture the secret. Additionally, an Attacker may determine the secret through off-line attacks on network traffic from an authentication attempt. Finally, an Attacker may be able to gain information about a Subscriber's Pre-registered Knowledge researching the subscriber or through other social engineering techniques. (For example, the subscriber might refer to his or her first pet in a conversation or blog.)

- *Something you are* may be replicated. An Attacker may obtain a copy of the token owner's fingerprint and construct a replica.

This document assumes that the Subscriber is not colluding with the Attacker who is attempting to falsely authenticate to the Verifier. With this assumption in mind, the threats to the token(s) used for e-authentication are listed in Table 4, along with some examples.

Table 4 – Token Threats

| Token Threats/Attacks | Description | Examples |
|---|---|---|
| Theft | A physical token is stolen by an Attacker. | A hardware cryptographic device is stolen. A One-time password device is stolen. A look-up secret token is stolen. A Cell phone is stolen. |
| Discovery | The responses to token prompts are easily discovered through searching various data sources. | The question "What high school did you attend?" is asked as a Pre-registered Knowledge Token, when the answer is commonly found on social media websites. |
| Duplication | The Subscriber's token has been copied with or without his or her knowledge. | Passwords written on paper are disclosed. Passwords stored in an electronic file are copied. Software PKI token (private key) copied. Look-up token copied. |
| Eavesdropping | The token secret or authenticator is revealed to the Attacker as the Subscriber is submitting the token to send over the network. | Passwords are learned by watching keyboard entry. Passwords are learned by Keystroke logging software. A PIN is captured from PIN pad device. |
| Offline cracking | The token is exposed using analytical methods outside the authentication mechanism. | A key is extracted by differential power analysis on stolen hardware cryptographic token. A software PKI token is subjected to dictionary attack to identify the correct password to use to decrypt the private key. |
| Phishing or pharming | The token secret or authenticator is captured by fooling the Subscriber into thinking the Attacker is a Verifier or RP. | A password is revealed by Subscriber to a website impersonating the Verifier. A password is revealed by a bank Subscriber in response to an email inquiry from a Phisher pretending to represent the bank. A password is revealed by the Subscriber at a bogus Verifier website reached through DNS re-routing. |
| Social engineering | The Attacker establishes a level of trust with a Subscriber in order to convince the Subscriber to reveal his or her token or token secret. | A password is revealed by the Subscriber to an officemate asking for the password on behalf of the Subscriber's boss. A password is revealed by a Subscriber in a telephone inquiry from an Attacker masquerading as a system administrator. |
| Online guessing | The Attacker connects to the Verifier online and attempts to guess a valid token authenticator in the context of that Verifier. | Online dictionary attacks are used to guess passwords. Online guessing is used to guess token authenticators for a one-time password token registered to a legitimate Claimant. |

### *6.2.1. Threat Mitigation Strategies*

Token related mechanisms that assist in mitigating the threats identified above are summarized in Table 5.

Table 5 - Mitigating Token Threats

| Token Threat/Attack | Threat Mitigation Mechanisms |
|---|---|
| Theft | - Use multi-factor tokens which need to be activated through a PIN or biometric. |
| Duplication | - Use tokens that are difficult to duplicate, such as hardware cryptographic tokens. |
| Discovery | - Use methods in which the responses to prompts cannot be easily discovered. |
| Eavesdropping | - Use tokens with dynamic authenticators where knowledge of one authenticator does not assist in deriving a subsequent authenticator.<br>- Use tokens that generate authenticators based on a token input value.<br>- Establish tokens through a separate channel. |
| Offline cracking | - Use a token with a high entropy token secret<br>- Use a token that locks up after a number of repeated failed activation attempts. |
| Phishing or pharming | - Use tokens with dynamic authenticators where knowledge of one authenticator does not assist in deriving a subsequent authenticator. |
| Social engineering | - Use tokens with dynamic authenticators where knowledge of one authenticator does not assist in deriving a subsequent authenticator. |
| Online guessing | - Use tokens that generate high entropy authenticators. |

There are several other strategies that may be applied to mitigate the threats described in Table 5:

- *Multiple factors* raise the threshold for successful attacks. If an Attacker needs to steal a cryptographic token and guess a password, then the work to discover both factors may be too high.

- *Physical security mechanisms* may be employed to protect a stolen token from duplication. Physical security mechanisms can provide tamper evidence, detection, and response.

- *Imposing password complexity rules* may reduce the likelihood of a successful guessing attack. Requiring the use of long passwords that don't appear in common dictionaries may force Attackers to try every possible password.

- *System and network security controls* may be employed to prevent an Attacker from gaining access to a system or installing malicious software.

- *Periodic training* may be performed to ensure the Subscriber understands when and how to report compromise (or suspicion of compromise) or otherwise recognize patterns of behavior that may signify an Attacker attempting to compromise the token.

- *Out of band techniques* may be employed to verify proof of possession of registered devices (e.g., cell phones).

### 6.3. Token Assurance Levels

This section discusses the requirements for tokens used at various levels of assurance.

#### 6.3.1. Requirements per Assurance Level

The following sections list token requirements for single and multi-token authentication.

##### 6.3.1.1. Single Token Authentication

Table 6 lists the assurance levels that may be achieved by each of the token types when used in a single-token authentication scheme. The requirements for each token are listed per assurance level. If token requirements are listed only at one assurance level, the token may be used at lower levels but shall satisfy the requirements given at whatever level is listed. If there is more than one box under "Verifier Requirements" for a given token type, it is only necessary to satisfy the requirements in one box.

Table 6 - Token Requirements Per Assurance Level

| Token Type | Level | Token Requirements | Verifier Requirements |
|---|---|---|---|
| Memorized Secret Token | Level 1 | The memorized secret may be a user chosen string consisting of 6 or more characters chosen from an alphabet of 90 or more characters, a randomly generated PIN consisting of 4 or more digits, or a secret with equivalent entropy.[20] | The Verifier shall implement a throttling mechanism that effectively limits the number of failed authentication attempts an Attacker can make on the Subscriber's account to 100 or fewer in any 30-day period.<br><br>Note: While an implementation that simply counted all failed authentication attempts in each calendar month and locked out the account when the limit was exceeded would technically meet the requirement, this is a poor choice for reasons of system availability. See Section 8.2.3 for more detailed advice. |
| | Level 2 | The memorized secret may be a randomly generated PIN consisting of 6 or more digits, a user generated string consisting of 8 or more characters chosen from an alphabet of 90 or more characters, or a secret with equivalent entropy.[20]<br><br>CSP implements dictionary or composition rule to constrain user-generated secrets. | The Verifier shall implement a throttling mechanism that effectively limits the number of failed authentication attempts an Attacker can make on the Subscriber's account to 100 or fewer in any 30-day period.<br><br>Note: While an implementation that simply counted all failed authentication attempts in each calendar month and locked out the account when the limit was exceeded would technically meet the requirement, this is a poor choice for reasons of system availability. See Section 8.2.3 for more detailed advice. |
| Pre-Registered Knowledge Token | Level 1 | The secret provides at least 14 bits of entropy.[20] | The Verifier shall implement a throttling mechanism that effectively limits the number of failed authentication attempts an Attacker can make on the Subscriber's account to 100 or fewer in any 30-day period.<br><br>Note: While an implementation that simply counted all failed authentication attempts in each calendar month and locked out the account when the limit was exceeded would technically meet the requirement, this is a poor choice for reasons of system availability. See Section 8.2.3 for more detailed advice. |

---

[20] For more information, see Table A.1 in Appendix A.

| Token Type | Level | Token Requirements | Verifier Requirements |
|---|---|---|---|
| | | The entropy in the secret cannot be directly calculated, e.g., user chosen or personal knowledge questions.<br><br>If the questions are not supplied by the user, the user shall select prompts from a set of at least five questions. | For these purposes, an empty answer is prohibited.<br><br>The Verifier shall verify the answers provided for at least three questions, and shall implement a throttling mechanism that effectively limits the number of failed authentication attempts an Attacker can make on the Subscriber's account to 100 or fewer in any 30-day period.<br><br>Note: While an implementation that simply counted all failed authentication attempts in each calendar month and locked out the account when the limit was exceeded would technically meet the requirement, this is a poor choice for reasons of system availability. See Section 8.2.3 for more detailed advice. |
| | Level 2 | The secret provides at least 20 bits of entropy.[20] | The Verifier shall implement a throttling mechanism that effectively limits the number of failed authentication attempts an Attacker can make on the Subscriber's account to 100 or fewer in any 30-day period.<br><br>Note: While an implementation that simply counted all failed authentication attempts in each calendar month and locked out the account when the limit was exceeded would technically meet the requirement, this is a poor choice for reasons of system availability. See Section 8.2.3 for more detailed advice. |
| | | The entropy in the secret cannot be directly calculated, e.g., user chosen or personal knowledge questions.<br><br>If the questions are not supplied by the user, the user shall select prompts from a set of at least seven questions. | For these purposes, an empty answer is prohibited.<br><br>The Verifier shall verify the answers provided for at least five questions, and shall implement a throttling mechanism that effectively limits the number of failed authentication attempts an Attacker can make on the Subscriber's account to 100 or fewer in any 30-day period.<br><br>Note: While an implementation that simply counted all failed authentication attempts in each calendar month and locked out the account when the limit was exceeded would technically meet the requirement, this is a poor |

| Token Type | Level | Token Requirements | Verifier Requirements |
|---|---|---|---|
| | | | choice for reasons of system availability. See Section 8.2.3 for more detailed advice. |
| Look-up Secret Token | Level 2 | The token authenticator has 64 bits of entropy.[20] | N/A |
| | | The token authenticator has at least 20 bits of entropy.[20] | The Verifier shall implement a throttling mechanism that effectively limits the number of failed authentication attempts an Attacker can make on the Subscriber's account to 100 or fewer in any 30-day period.<br><br>Note: While an implementation that simply counted all failed authentication attempts in each calendar month and locked out the account when the limit was exceeded would technically meet the requirement, this is a poor choice for reasons of system availability. See Section 8.2.3 for more detailed advice. |
| Out of Band Token | Level 2 | The token is uniquely addressable and supports communication over a channel that is separate from the primary channel for e-authentication. | The Verifier generated secret shall have at least 64 bits of entropy.[20] |
| | | | The Verifier generated secret shall have at least 20 bits of entropy[20] and the Verifier shall implement a throttling mechanism that effectively limits the number of failed authentication attempts an Attacker can make on the Subscriber's account to 100 or fewer in any 30-day period.<br><br>Note: While an implementation that simply counted all failed authentication attempts in each calendar month and locked out the account when the limit was exceeded would technically meet the requirement, this is a poor choice for reasons of system availability. See Section 8.2.3 for more detailed advice. |
| SF One-Time Password Device | Level 2 | Shall use Approved block cipher or hash function to combine a symmetric key stored on device with a nonce to generate a one-time password.<br><br>The nonce may be a date and time, or a counter generated on the device. | The one-time password shall have a limited lifetime, on the order of minutes.<br><br>The cryptographic module performing the verifier function shall be validated at FIPS 140-2 Level 1 or higher.[21] |
| SF Cryptographic | Level 2 | The cryptographic module shall be validated at FIPS 140-2 Level 1 or | Verifier generated token input (e.g., a nonce or challenge) has at |

---

[21] Products validated under subsequent versions of FIPS 140-2 are also acceptable.

| Token Type | Level | Token Requirements | Verifier Requirements |
|---|---|---|---|
| Device | | higher.[21] | least 64 bits of entropy.[20] |
| MF Software Cryptographic Token | Level 3 | The cryptographic module shall be validated at FIPS 140-2 Level 1 or higher.[21] Each authentication shall require entry of the password or other activation data and the unencrypted copy of the authentication key shall be erased after each authentication. | Verifier generated token input (e.g., a nonce or challenge) has at least 64 bits of entropy.[20] |
| MF OTP Hardware Token | Level 4 | Cryptographic module shall be FIPS 140-2 validated Level 2 or higher; with physical security at FIPS 140-2 Level 3 or higher.[21]<br><br>The one-time password shall be generated by using an Approved block cipher or hash function to combine a symmetric key stored on a personal hardware device with a nonce to generate a one-time password.<br><br>The nonce may be a date and time, a counter generated on the device. Each authentication shall require entry of a password or other activation data through an integrated input mechanism. | The one-time password shall have a limited lifetime of less than 2 minutes. |
| MF Hardware Cryptographic Token | Level 4 | Cryptographic module shall be FIPS 140-2 validated, Level 2 or higher; with physical security at FIPS 140-2 Level 3 or higher.[21] Shall require the entry of a password, PIN, or biometric to activate the authentication key. Shall not allow the export of authentication keys. | Verifier generated token input (e.g., a nonce or challenge) has at least 64 bits of entropy.[20] |

### 6.3.1.2. Multi-Token Authentication

When two of the token types are combined for a multi-token authentication scheme, Table 7 shows the highest possible assurance level that can be achieved by the combination.[22]

---

[22] Note that the table displays tokens that exhibit the properties of "something you have" and "something you know".

Table 7 - Assurance Levels for Multi-Token E-Authentication Schemes[23]

| | Memorized Secret Token | Pre-registered Knowledge Token | Look-up Secret Token | Out of Band Token | SF OTP Device | SF Crypto-graphic Device | MF Software Cryptographic Token | MF OTP Device | MF Crypto-graphic Device |
|---|---|---|---|---|---|---|---|---|---|
| Memorized Secret Token | Level 2 | Level 2 | Level 3 | Level 3 | Level 3 | Level 3 | Level 3 | Level 4 | Level 4 |
| Pre-registered Knowledge Token | x | Level 2 | Level 3 | Level 3 | Level 3 | Level 3 | Level 3 | Level 4 | Level 4 |
| Look-up Secret Token | x | x | Level 2 | Level 3 | Level 3 | Level 3 | Level 3 | Level 4 | Level 4 |
| Out of Band Token | x | x | x | Level 2 | Level 2 | Level 2 | Level 3 | Level 4 | Level 4 |
| SF OTP Device | x | x | x | x | Level 2 | Level 2 | Level 3 | Level 4 | Level 4 |
| SF Cryptographic Device | x | x | x | x | x | Level 2 | Level 3 | Level 4 | Level 4 |
| MF Software Cryptographic Token | x | x | x | x | x | x | Level 3 | Level 4 | Level 4 |
| MF OTP Device | x | x | x | x | x | x | x | Level 4 | Level 4 |
| MF Cryptographic Device | x | x | x | x | x | x | x | x | Level 4 |

[23] The boxes marked with an "x" denote that the combination already appears in the table

The principles used in generating Table 7 are as follows. Level 3 can be achieved using two tokens rated at Level 2 that represent two different factors of authentication. Since this specification does not recognize biometrics as a stand-alone token for remote authentication, achieving Level 3 with separate Level 2 tokens implies *something you have* and *something you know*:

Token (Level 2, *something you have*) + Token (Level 2, *something you know*) → Token(Level 3)

In all other cases, combinations of tokens are considered to achieve the Level of the highest rated token.

For example, a Memorized Secret Token combined with a Look-up Secret Token can be used to achieve Level 3 authentication, since the look-up secret token is "something you have" and the Memorized Secret Token is "something you know". However, combining a MF software cryptographic token (which is rated at Level 3) and a Memorized Secret Token (which is rated at Level 2) achieves an overall level of 3, since the addition of the Memorized Secret Token does not increase the assurance of the combination.

It should be noted that to achieve Level 4 with a single token or token combination, one of the tokens needs to be usable with an authentication process that strongly resists man-in-the-middle attacks. While it is possible to meet this requirement with a wide variety of token types, certain choices of tokens may complicate the task of designing a protocol that meets Level 4 requirements for authentication process (as described in Section 8 of this document). In particular, one-time password devices that rely exclusively on the human user for input and output may be especially problematic and may need to be supplemented with a software cryptographic token to provide strong man-in-the-middle resistance.

# 7. Token and Credential Management

## 7.1. Overview

As introduced in Section 4, credentials are objects that bind identity to a token. To maintain the level of assurance provided by an e-authentication solution, credentials and tokens shall be managed to reflect any changes in that binding. This section discusses token and credential management activities performed by the CSP subsequent to the registration, identity proofing and issuance activities described in Section 5. This includes the lifecycle management activities for the token and credential. The activities that must be performed by the CSP depend in part upon the nature of the credentials and the token itself.

### 7.1.1. Categorizing Credentials

This specification categorizes credentials according to two orthogonal perspectives. Some classes of credentials can be distributed to relying parties, while others cannot be disclosed by the CSP without compromising the token itself. Another classification indicates whether the binding represented in the credential is tamper-evident.

Credentials that describe the binding in a way that does not compromise the token are referred to as *Public Credentials*. The classic example of a Public Credential is a public key certificate; it is mathematically infeasible to calculate the user's private key even with knowledge of the corresponding public key. Credentials that cannot be disclosed by the CSP because the contents can be used to compromise the token are considered *Private Credentials*. The classic example of a Private Credential is the hashed value of a password, since this hash can be used to perform an offline attack on the password.

Credentials that describe the binding between a user and token in a tamper-evident fashion are considered *Strongly Bound Credentials*. For example, modification of a digitally signed credential (such as a public key certificate) can be easily detected through signature verification. The binding between a user and token can be modified in *Weakly Bound Credentials* without invalidating the credentials. Weakly bound credentials require supplemental integrity protection and/or access controls to ensure that the binding represented by the credential remains accurate. For example, replacing the value of a hashed password in a password file associates the user with a new password, so access to this file is restricted to system users and processes.

Strongly bound credential mechanisms require little or no additional integrity protection, whereas weakly bound credentials require additional integrity protection or access controls to ensure that unauthorized parties cannot spoof or tamper with the binding of the identity to the token representation within the credential.

Unencrypted password files are private credentials that are weakly bound, and hence need to be afforded confidentiality as well as integrity protection. Signed password files are private credentials that are strongly bound and therefore require confidentiality protection but no additional integrity protection. An unsigned pairing of a public key and the name of its owner or a self-signed certificate is an example of a public credential that is weakly bound. Finally, a CA-signed public key certificate represents a public credential that is strongly bound.

CSPs and Verifiers are trusted to obey the requirements in this section as well as Section 8.

### 7.1.2. Token and Credential Management Activities

The CSP manages tokens and credentials. The RA establishes the Applicant's identity, and the CSP is responsible for generating credentials and supplying the Subscriber with a token or allowing the Subscriber to register his or her own token as described in Section 5. The CSP is responsible for some or all of the following token and credential management activities following issuance of the token and credential:

- *Credential storage* – After the credential has been created, the CSP may be responsible for maintaining the credentials in storage. In cases where the credentials are stored by the CSP, the level of security afforded to the credential will depend on the type of credential issued. For private credentials, additional confidentiality mechanisms are required in storage, whereas for public credentials, this is not necessary. Similarly, for weakly bound credentials, additional integrity protection is needed in storage, unlike strongly bound credentials. Finally, credentials need to be available to allow CSPs and Verifiers to determine the identity of the corresponding token owner.

- *Token and credential verification services* – In many e-authentication scenarios, the Verifier and the CSP are not part of the same entity. In these cases, the CSP is responsible for providing the Verifier with the information needed to facilitate the token and credential verification process. The CSP may provide token and credential verification services to Verifiers. For example, the Verifier may request the CSP to verify the password submitted by the Claimant against the CSP's local password database.

- *Token and credential renewal /re-issuance* – Certain types of tokens and credentials may support the process of renewal or re-issuance. During renewal, the usage or validity period of the token and credential is extended without changing the Subscriber's identity or token. During re-issuance, a new credential is created for a Subscriber with a new identity and/or a new token.

    The CSP establishes suitable policies for renewal and re-issuance of tokens and credentials. The CSP may establish a time period prior to the expiration of the credential, when the Subscriber can request renewal or re-issuance following successful authentication using his or her existing, unexpired token and credential. For example, a CSP may allow a digital certificate to be renewed for another year prior to the expiry of the current certificate by proving possession and control of the existing token (i.e., the private key).

Once the Subscriber's credentials have expired, the Subscriber may be required to re-establish his or her identity with the CSP; this is typically the case with CSPs that issue digital certificates. Conversely, the CSP may establish a grace period for the renewal or re-issuance of an expired credential, such that the Subscriber can request renewal/re-issuance of his or her credential even after it has expired without the need to re-establish his or her identity with the CSP. For example, if a Claimant attempts to login to a username/password based system on which his or her password has already expired, and the system supports a grace period, the user may be prompted to create a new password and supply the last password for verification purposes. The use of expired tokens or credentials to invoke renewal/re-issuance is more practical when the Verifier and CSP are part of the same entity.

The public key certificate for a Subscriber may be renewed with the same public key, or may be re-issued with a new public key. Passwords are seldom renewed so that the life of the existing password is extended for another period. Usually the account name/password credential for a Subscriber is renewed by having the Subscriber select a new password.

- *Token and credential revocation and destruction* – The CSP is responsible for maintaining the revocation status of credentials and destroying the credential at the end of its life. Explicit and elaborate revocation mechanisms may be required for "public credentials" since these credentials are disseminated widely, possibly with a preset validity period. For example, public key certificates are revoked using Certificate Revocation Lists (CRLs) after the certificates are distributed.

    "Private credentials" are held closely by the CSP, and hence the revocation and destruction of these credentials is implemented easily through an update of the CSP's local credential stores. Credentials that bind usernames/passwords are instantaneously revoked and destroyed if the CSP deletes its mapping between the username and the password. Certain types of tokens may need to be explicitly deleted or zeroized at the end of the credential life in order to permanently disable the token and prevent its unauthorized reuse. For example, a Multi-factor Hardware Cryptographic Token may need to be zeroized to ensure that all of the information pertaining to the Subscriber is deleted from the token.

    The CSP may be responsible for ensuring that hardware tokens are collected and cleared of any data when the Subscriber no longer has a need for its use. The CSP may establish policies for token collection to avoid the possibility of unauthorized use of the token after it is considered out of use. The CSP may destroy such collected tokens, or zeroize them to ensure that there are no remnants of information that can be used by an Attacker to derive the token value. For example, a Subscriber who is issued a hardware OTP token by a CSP may be required by policy to return the token to the CSP at the end of its life, or when the Subscriber's association with that CSP terminates.

- *Records retention* – The CSP or its representative is responsible for maintaining a record of the registration, history, and status of each token and credential, including revocation. CSPs operated by or on behalf of executive branch agencies shall also follow either the General Records Schedule established by the National Archives and Records Administration or an agency-specific schedule as applicable. All other entities shall comply with their respective records retention policies in accordance with whatever laws apply to those entities. A minimum record retention period is required at Level 2 and above.
- *Security controls* – The CSP is responsible for implementing and maintaining appropriate security controls contained in NIST SP 800-53. The security control baseline for CSPs is specified in terms of a FIPS 200 impact level for each assurance level. (See Section 7.3, below.)

### 7.2. Token and Credential Management Threats

Tokens and credentials can only be as strong as the strength of the management mechanisms used to secure them. The CSP is responsible for mitigating threats to the management operations described in the last section. Token and credential management threats are described below; they are categorized in accordance with the management activity to which they apply.

These threats represent the potential to breach the confidentiality, integrity and availability of tokens and credentials during the CSP activities, and are listed below.

Table 8 - Token and Credential Management Threats

| Token and Credential Management Activity | Threat/Attack | Example |
|---|---|---|
| Credential storage | Disclosure | Usernames and passwords stored in a system file are revealed. |
| | Tampering | The file that maps usernames to passwords within the CSP is hacked so that the mappings are modified, and existing passwords are replaced by passwords known to the Attacker. |
| Token and credential verification services | Disclosure | An Attacker is able to view requests and responses between the CSP and the Verifier. |
| | Tampering | An Attacker is able to masquerade as the CSP and provide bogus responses to the Verifier's password verification requests. |
| | Unavailability | The password file or the CSP is unavailable to provide password and username mappings. |
| | | Public key certificates for Claimants are unavailable to the Verifier because the directory systems are down (for example for maintenance or as a result of a denial of service attack). |
| Token and credential issuance/renewal/re-issuance | Disclosure | Password renewed by the CSP for a Subscriber is copied by an Attacker as it is transported from the CSP to the Subscriber. |
| | Tampering | New password created by the Subscriber is modified by an Attacker as it is being submitted to the CSP to replace an expired password. |
| | Unauthorized issuance | The CSP is compromised through unauthorized physical or logical access resulting in issuance of fraudulent credentials. |

| Token and Credential Management Activity | Threat/Attack | Example |
|---|---|---|
| | Unauthorized renewal/re-issuance | Attacker fools the CSP into re-issuing the credential for a current Subscriber – the new credential binds the current Subscriber's identity with a token provided by the Attacker. |
| | | Attacker is able to take advantage of a weak credential renewal protocol to extend the credential validity period for a current Subscriber. |
| Token and credential revocation/destruction | Delayed revocation/destruction of credentials | Stale CRLs allow accounts (that should have been locked as a result of credential revocation) to be used by an Attacker. |
| | | User accounts are not deleted when employees leave a company leading to a possible use of the old accounts by unauthorized persons. |
| | Token use after decommissioning | A hardware token is used after the corresponding credential was revoked or expired. |

### 7.2.1. Threat Mitigation Strategies

Token and credential management related mechanisms that assist in mitigating the threats identified above are summarized in Table 9.

## 7.3. Token and Credential Management Assurance Levels

### 7.3.1. Requirements per Assurance Level

The stipulations for management of tokens and credentials by the CSP and Verifier are described below for each assurance level. The stipulations described at each level in this section are incremental in nature; requirements stipulated at lower levels are implicitly included at higher levels.

Table 9 - Token and Credential Threat Mitigation Strategies

| Token and Credential Management Activity | Threat/Attack | Mitigation Strategy |
|---|---|---|
| Credential storage | Disclosure | Use access control mechanisms that protect against unauthorized disclosure of credentials held in storage. |
| | Tampering | Use access control mechanisms that protect against unauthorized tampering of credentials and tokens. |
| Token and credential verification services | Disclosure | Use a communication protocol that offers confidentiality protection. |
| | Tampering | Ensure that Verifiers authenticate the CSP prior to accepting a verification response from that CSP. |
| | | Use a communication protocol that offers integrity protection. |
| | Unavailability | Ensure that the CSP has a well developed and tested Contingency Plan. |
| Token and credential issuance/renewal/re-issuance | Disclosure | Use a communication protocol that provides confidentiality protection of session data. |
| | Tampering | Use a communication protocol that allows the Subscriber to authenticate the CSP prior to engaging in token re-issuance activities and protects the integrity of the data passed. |
| | Unauthorized issuance | Implement physical and logical access controls to prevent compromise of the CSP. See [FISMA] for details on security controls. |
| | Unauthorized renewal/re-issuance | Establish policy that Subscriber shall prove possession of the old token to successfully negotiate the re-issuance process. Any attempt to negotiate the re-issuance process using an expired or revoked token should fail. |
| Credential revocation/destruction | Delayed revocation/destruction of credentials | Revoke/Destroy credentials as soon as notification that the credentials should be revoked or destroyed. |
| | Token use after decommissioning | Destroy tokens after their corresponding credentials have been revoked. |

#### 7.3.1.1. Level 1

At Level 1, the following shall be required:

- *Credential storage* – Files of shared secrets used by Verifiers at Level 1 authentication shall be protected by access controls that limit access to administrators and only to those applications that require access. Such shared secret files shall not contain the plaintext passwords; typically they contain a one-way hash or "inversion" of the password. In addition, any method allowed for the protection of long-term shared secrets at Level 2 or above may be used at Level 1.

- *Token and credential verification services* – Long term token secrets should not be shared with other parties unless absolutely necessary.

- *Token and credential renewal / re-issuance* – No stipulation

- *Token and credential revocation and destruction* – No stipulation

- *Records retention* – No stipulation

- *Security controls* – No stipulation

#### 7.3.1.2. Level 2

At Level 2, the following shall be required:

- *Credential storage* – Files of shared secrets used by CSPs at Level 2 shall be protected by access controls that limit access to administrators and only to those applications that require access. Such shared secret files shall not contain the plaintext passwords or secrets; two alternative methods may be used to protect the shared secret:

    1. Passwords may be concatenated to a variable salt (variable across a group of passwords that are stored together) and then hashed with an Approved algorithm so that the computations used to conduct a dictionary or exhaustion attack on a stolen password file are not useful to attack other similar password files. The hashed passwords are then stored in the password file. The variable salt may be composed using a global salt (common to a group of passwords) and the username (unique per password) or some other technique to ensure uniqueness of the salt within the group of passwords.

    2. Shared secrets may be encrypted and stored using Approved encryption algorithms and modes, and the needed secret decrypted only when immediately required for authentication. In addition, any method allowed to protect shared secrets at Level 3 or 4 may be used at Level 2.

- *Token and credential verification services* – Long term shared authentication secrets, if used, shall never be revealed to any other party except Verifiers operated by the CSP; however, session (temporary) shared secrets may be provided by the CSP to independent Verifiers.

  Cryptographic protections are required for all messages between the CSP and Verifier which contain private credentials or assert the validity of weakly bound or potentially revoked credentials. Private credentials shall only be sent through a protected session to an authenticated party to ensure confidentiality and tamper protection.

  The CSP may send the Verifier a message that either asserts that a weakly bound credential is valid, or that a strongly bound credential has not been subsequently revoked. In this case, the message shall be logically bound to the credential, and the message, the logical binding, and the credential shall all be transmitted within a single integrity protected session between the Verifier and the authenticated CSP. If revocation is an issue, the integrity protected messages shall either be time stamped, or the session keys shall expire with an expiration time no longer than that of the revocation list. Alternatively, the time stamped message, binding, and credential may all be signed by the CSP, although, in this case, the three in combination would comprise a strongly bound credential with no need for revocation.

- *Token and credential renewal/re-issuance* – The CSP shall establish suitable policies for renewal and re-issuance of tokens and credentials. Proof-of-possession of the unexpired current token shall be demonstrated by the Claimant prior to the CSP allowing renewal and re-issuance. Passwords shall not be renewed; they shall be re-issued. After expiry of current token and any grace period, renewal and re-issuance shall not be allowed. Upon re-issuance, token secrets shall not be set to a default or reused in any manner. All interactions shall occur over a protected session such as SSL/TLS.

- *Token and credential revocation and destruction* – CSPs shall revoke or destroy credentials and tokens within 72 hours after being notified that a credential is no longer valid or a token is compromised to ensure that a Claimant using the token cannot successfully be authenticated. If the CSP issues credentials that expire automatically within 72 hours (e.g., issues fresh certificates with a 24 hour validity period each day) then the CSP is not required to provide an explicit mechanism to revoke the credentials. CSPs that register passwords shall ensure that the revocation or de-registration of the password can be accomplished in no more than 72 hours. CAs cross-certified with the Federal Bridge CA at the Citizen and Commerce Class Basic, Medium and High or Common Certificate Policy levels are considered to meet credential status and revocation provisions of this level.

- *Records retention* – A record of the registration, history, and status of each token and credential (including revocation) shall be maintained by the CSP or its representative. The record retention period of data for Level 2 credentials is seven years and six months beyond the expiration or revocation (whichever is later) of the credential. CSPs operated by or on behalf of executive branch agencies shall also follow either the General Records Schedule established by the National Archives and Records Administration or an agency-specific schedule as applicable. All other entities shall comply with their respective records retention policies in accordance with whatever laws apply to those entities.

- *Security controls* – The CSP must employ appropriately tailored security controls from the low baseline of security controls defined in [SP 800-53] and must ensure that the minimum assurance requirements associated with the low baseline are satisfied.

### 7.3.1.3. Level 3

At Level 3, the following is required:

- *Credential storage*[24] – Files of long-term shared secrets used by CSPs or Verifiers at Level 3 shall be protected by access controls that limit access to administrators and only to those applications that require access. Such shared secret files shall be encrypted so that:

  1. The encryption key for the shared secret file is encrypted under a key held in a FIPS 140-2 Level 2 or higher validated hardware cryptographic module or any FIPS 140-2 Level 3 or 4 cryptographic module and decrypted only as immediately required for an authentication operation.

  2. Shared secrets are protected as a key within the boundary of a FIPS 140-2 Level 2 or higher validated hardware cryptographic module or any FIPS 140-2 Level 3 or 4 cryptographic module and is not exported in plaintext from the module.

  Strongly bound credentials support tamper detection mechanisms such as digital signatures, but weakly bound credentials can be protected against tampering using access control mechanisms as described above.

- *Token and credential verification services* – CSPs shall provide a secure mechanism to allow Verifiers or RPs to ensure that the credentials are valid. Such mechanisms may include on-line validation servers or the involvement of CSP servers that have access to status records in authentication transactions.

  Temporary session authentication keys may be generated from long-term shared secret keys by CSPs and distributed to third party Verifiers, as a part of the verification services offered by the CSP, but long-term shared secrets shall not be shared with any third parties, including third party Verifiers. This type

---

[24] With regard to references to FIPS 140-2, products validated under subsequent versions of FIPS 140-2 are also acceptable.

of third-party (or delegated) verification is used in the realm of GSM (Global System for Mobile Communications) roaming; the locally available network authenticates the "roaming" Subscriber using a temporary session authentication key received from the Base Station. Such temporary session authentication keys are typically created by cryptographically combining the long term shared secret with a nonce challenge, to generate a session key. The challenge and session key are securely transmitted to the Verifier. The Verifier in turn sends only the challenge to the Claimant, and the Claimant applies the challenge to the long-term shared secret to generate the session key. Both Claimant and Verifier now share a session key, which can be used for authentication. Such verification schemes are permitted at this level provided that Approved cryptographic algorithms are used for all operations.

Token and credential verification services categorized as FIPS 199 "Moderate" or "High" for availability shall be protected in accordance with the Contingency Planning (CP) controls specified in NIST SP 800-53 to provide an adequate level of availability needed for the service.

- *Token and credential renewal /re-issuance* – Renewal and re-issuance shall only occur prior to expiration of the current credential. Claimants shall authenticate to the CSP using the existing token and credential in order to renew or re-issue the credential. All interactions shall occur over a protected session such as SSL/TLS.

- *Credential revocation and destruction* – CSPs shall have a procedure to revoke credentials and tokens within 24 hours. The certificate status provisions of CAs cross-certified with the Federal Bridge CA at the Basic, Medium, High or Common Certificate Policy levels are considered to meet credential status and revocation provisions of this level. Verifiers shall ensure that the tokens they rely upon are either freshly issued (within 24 hours) or still valid. Shared secret based authentication systems may simply remove revoked Subscribers from the verification database.

- *Records retention* – All stipulations from Level 2 apply.

- *Security controls* – The CSP must employ appropriately tailored security controls from the moderate baseline of security controls defined in [SP 800-53] and must ensure that the minimum assurance requirements associated with the moderate baseline are satisfied.

### 7.3.1.4. Level 4

At Level 4, the following is required:

- *Credential storage* – No additional stipulation.
- *Token and credential verification services* – No additional stipulation.

- *Token and credential renewal/re-issuance* – Sensitive data transfers shall be cryptographically authenticated using keys bound to the authentication process. All temporary or short-term keys derived during the original authentication operation shall expire and re-authentication shall be required after not more than 24 hours from the initial authentication.

- *Token and credential revocation and destruction* – CSPs shall have a procedure to revoke credentials within 24 hours. Verifiers or RPs shall ensure that the credentials they rely upon are either freshly issued (within 24 hours) or still valid. The certificate status provisions of CAs cross-certified with the Federal Bridge CA at the High and Common Certificate Policies shall be considered to meet credential status provisions of Level 4. [FBCA1]

  It is generally good practice to destroy a token within 48 hours of the end of its life or the end of the Subscriber's association with the CSP. Destroying includes either the physical destruction of the token or cleansing it of all information related to the Subscriber.

- *Records retention* – All stipulations from Levels 2 and 3 apply. The minimum record retention period for Level 4 credential data is ten years and six months beyond the expiration or revocation of the credential.

- *Security controls* – The CSP must employ appropriately tailored security controls from the moderate baseline of security controls defined in [SP 800-53] and must ensure that the minimum assurance requirements associated with the moderate baseline are satisfied.

### 7.3.2. Relationship of PKI Policies to E-Authentication Assurance Levels

Appendix B specifies the mapping between the Federal PKI Certificate Policies and the requirements in Section 7.

# 8. Authentication Process

## 8.1. Overview

The authentication process establishes the identity of the Claimant to the Verifier with a certain degree of assurance. It is implemented through an authentication protocol message exchange, as well as management mechanisms at each end that further constrain or secure the authentication activity. One or more of the messages of the authentication protocol may need to be carried on a protected session. This is illustrated in Figure 3.

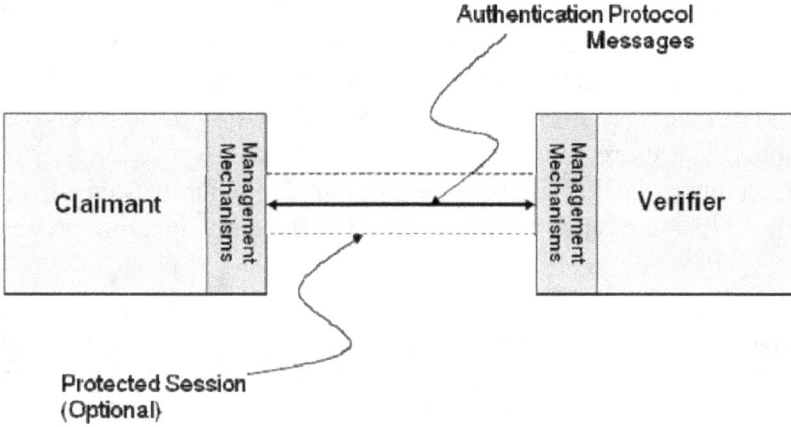

Figure 3 - *Authentication Process Model*

*An authentication protocol is a defined sequence of messages between a Claimant and a Verifier that demonstrates that the Claimant has control of a valid token to establish his or her identity, and optionally, demonstrates to the Claimant that he or she is communicating with the intended Verifier.* An exchange of messages between a Claimant and a Verifier that results in authentication (or authentication failure) between the two parties is an authentication protocol run. During or after a successful authentication protocol run, a protected communication session may be created between the two parties; this protected session may be used to exchange the remaining messages of the authentication protocol run, or to exchange session data between the two parties.

Management mechanisms may be implemented on the Claimant and the Verifier to further enhance the authentication process. For example, trust anchors may be established at the Claimant to enable the authentication of the Verifier using public key mechanisms such as TLS. Similarly, mechanisms may be implemented on the Verifier to limit the rate of online guessing of passwords by an Attacker who is trying to authenticate as a legitimate Claimant. Further, detection of authentication transactions originating from an unexpected location or channel for a Claimant, or indicating use of an unexpected hardware or software configuration, may indicate increased risk levels and motivate additional confirmation of the Claimant's identity.

At the conclusion of the authentication protocol run, the verifier might issue a secondary authentication credential, such as a cookie, to the Claimant and rely upon it to authenticate the claimant in the near future. Requirements for doing this securely are in Section 9.

## 8.2. Authentication Process Threats

In general, attacks that reveal long-term token secrets are worse than attacks that reveal short-term authentication secrets or session data, because in the former, the Attacker can then use the token secret to assume a Subscriber's identity and do greater harm.

RAs, CSPs, and Verifiers are ordinarily trustworthy (in the sense of being correctly implemented and not deliberately malicious). However, Claimants or their systems may not be trustworthy (or else their identity claims could simply be trusted). Moreover, while RAs, CSPs, and Verifiers are normally trustworthy, they are not invulnerable, and could become corrupted. Therefore, authentication protocols that expose long-term authentication secrets more than is absolutely required, even to trusted entities, should be avoided. Table 10 lists the types of threats posed to the authentication process.

Table 10 - Authentication Process Threats

| Type of Attack | Description | Example |
|---|---|---|
| Online guessing | An Attacker performs repeated logon trials by guessing possible values of the token authenticator. | An Attacker navigates to a web page and attempts to log in using a Subscriber's username and commonly used passwords, such as "password" and "secret". |
| Phishing | A Subscriber is lured to interact with a counterfeit Verifier, and tricked into revealing his or her token secret, sensitive personal data or authenticator values that can be used to masquerade as the Subscriber to the Verifier. | A Subscriber is sent an email that redirects him or her to a fraudulent website and is asked to log in using his or her username and password. |
| Pharming | A Subscriber who is attempting to connect to a legitimate Verifier, is routed to an Attacker's website through manipulation of the domain name service or routing tables. | A Subscriber is directed to a counterfeit website through DNS poisoning, and reveals or uses his or her token believing he or she is interacting with the legitimate Verifier. |
| Eavesdropping | An Attacker listens passively to the authentication protocol to capture information which can be used in a subsequent active attack to masquerade as the Claimant. | An Attacker captures the transmission of a password or password hash from a Claimant to a Verifier. |
| Replay | An Attacker is able to replay previously captured messages (between a legitimate Claimant and a Verifier) to authenticate as that Claimant to the Verifier. | An Attacker captures a Claimant's password or password hash from an actual authentication session, and replays it to the Verifier to gain access at a later time. |

| Type of Attack | Description | Example |
|---|---|---|
| Session hijack | An Attacker is able to insert himself or herself between a Subscriber and a Verifier subsequent to a successful authentication exchange between the latter two parties. The Attacker is able to pose as a Subscriber to the Verifier/RP or vice versa to control session data exchange. | An Attacker is able to take over an already authenticated session by eavesdropping on or predicting the value of authentication cookies used to mark HTTP requests sent by the Subscriber. |
| Man-in-the-middle | The Attacker positions himself or herself in between the Claimant and Verifier so that he or she can intercept and alter the content of the authentication protocol messages. The Attacker typically impersonates the Verifier to the Claimant and simultaneously impersonates the Claimant to the Verifier. Conducting an active exchange with both parties simultaneously may allow the Attacker to use authentication messages sent by one legitimate party to successfully authenticate to the other. | An Attacker breaks into a router that forwards messages between the Verifier and a Claimant. When forwarding messages, the Attacker substitutes his or her own public key for that of the Verifier. The Claimant is tricked into encrypting his or her password so that the Attacker can decrypt it. |
| | | An Attacker sets up a fraudulent website impersonating the Verifier. When an unwary Claimant tries to log in using his or her one-time password device, the Attacker's website simultaneously uses the Claimant's one-time password to log in to the real Verifier. |

### 8.2.1. Other Threats

Attacks are not limited to the authentication protocol itself. Other attacks include:

- Denial of Service attacks in which the Attacker overwhelms the Verifier by flooding it with a large amount of traffic over the authentication protocol;

- Malicious code attacks that may compromise or otherwise exploit authentication tokens;

- Attacks that fool Claimants into using an insecure protocol, when the Claimant thinks that he or she is using a secure protocol, or trick the Claimant into overriding security controls (for example, by accepting server certificates that cannot be validated).

The purpose of flooding attacks is to overwhelm the resources used to support an authentication protocol to the point where legitimate Claimants cannot reach the Verifier or to slow down the process to make it more difficult for the Claimant to reach the Verifier. For example, a Verifier that implements an authentication protocol that uses encryption/decryption is sent a large number of protocol messages causing the Verifier to be crippled due to the use of excessive system resources to encrypt/decrypt. Nearly all authentication protocols are susceptible to flooding attacks; possible ways to resist such attacks is through the use of distributed Verifier architectures, use of load balancing techniques to distribute protocol requests to multiple mirrored Verifier systems, or other similar techniques.

Malicious code could be introduced into the Claimant's computer system for the purpose of compromising or otherwise exploiting the Claimant's token. The malicious code may be introduced by many means, including the threats detailed below. There are many countermeasures (e.g., virus checkers and firewalls) that can mitigate the risk of malicious code on Claimant systems. General good practice to mitigate malicious code threats is outside the scope of this document[25]. Hardware tokens prevent malicious software from extracting and copying the token secret. However, malicious code may still misuse the token, particularly if activation data is presented to the token via the computer.

### 8.2.2. Threat Mitigation Strategies

The following are strategies that can be incorporated in authentication processes to mitigate the attacks listed in the previous section:

- *Online guessing resistance* – An authentication process is resistant to online guessing attacks if it is impractical for the Attacker, with no a priori knowledge of the token authenticator, to authenticate successfully by repeated authentication attempts with guessed authenticators. The entropy of the authenticator, the nature of the authentication protocol messages, and other management mechanisms at the Verifier contribute to this property. For example, password authentication systems can make targeted password guessing impractical by requiring use of high-entropy passwords and limiting the number of unsuccessful authentication attempts, or by controlling the rate at which attempts can be carried out. (See Appendix A and Table 6 in Section 6.3.1.). Similarly, to resist untargeted password attacks, a Verifier may supplement these controls with network security controls.

- *Phishing and pharming resistance (verifier impersonation)* – An authentication process is resistant to phishing and pharming (also known as Verifier impersonation,) if the impersonator does not learn the value of a token secret or a token authenticator that can be used to act as a Subscriber to the genuine Verifier. In the most general sense, this assurance can be provided by the same mechanisms that provide the strong man-in-the-middle resistance described later in this section; however, long term secrets can be protected against phishing and pharming simply by the use of a tamper resistant token, provided that the long term secret cannot be reconstructed from a Token Authenticator. To decrease the likelihood of phishing and pharming attacks, it

---

[25] See SP 800-53, *Recommended Security Controls For Federal Information Systems*

is recommended that the Claimant authenticate the Verifier using cryptographic mechanisms prior to submitting the token authenticator to the supposed Verifier. Additionally, management mechanisms can be implemented at the Verifier to send a Claimant personalized content after successful authentication of the Claimant or the Claimant's device. (Refer to Section 8.2.4 for further details on personalization.) This allows the Claimant to achieve a higher degree of assurance of the authenticity of the Verifier before proceeding with the remainder of the session with the Verifier or RP. It should be mentioned, however, that there is no foolproof way to prevent the Claimant from revealing any sensitive information to which he or she has access.

- *Eavesdropping resistance* – An authentication process is resistant to eavesdropping attacks if an eavesdropper who records all the messages passing between a Claimant and a Verifier finds it impractical to learn the Claimant's token secret or to otherwise obtain information that would allow the eavesdropper to impersonate the Subscriber in a future authentication session. Eavesdropping-resistant protocols make it impractical[26] for an Attacker to carry out an off-line attack where he or she records an authentication protocol run and then analyzes it on his or her own system for an extended period to determine the token secret or possible token authenticators. For example, an Attacker who captures the messages of a password-based authentication protocol run may try to crack the password by systematically trying every password in a large dictionary, and comparing it with the protocol run data. Protected session protocols, such as TLS, provide eavesdropping resistance.

- *Replay resistance* – An authentication process resists replay attacks if it is impractical to achieve a successful authentication by recording and replaying a previous authentication message. Protocols that use nonces or challenges to prove the "freshness" of the transaction are resistant to replay attacks since the Verifier will easily detect that the old protocol messages replayed do not contain the appropriate nonces or timeliness data related to the current authentication session.

- *Hijacking resistance* – An authentication process and data transfer protocol combination are resistant to hijacking if the authentication is bound to the data transfer in a manner that prevents an adversary from participating actively in the data transfer session between the Subscriber and the Verifier or RP without being detected. This is a property of the relationship of the authentication protocol and the subsequent session protocol used to transfer data. This binding is usually accomplished by generating a per-session shared secret during the authentication process that is subsequently used by the

---

[26] "Impractical" is used here in the cryptographic sense of nearly impossible, that is there is always a small chance of success, but even the Attacker with vast resources will nearly always fail. For off-line attacks, impractical means that the amount of work required to "break" the protocol is at least on the order of $2^{80}$ cryptographic operations. For on-line attacks impractical means that the number of possible on-line trials is very small compared to the number of possible key or password values.

Subscriber and the Verifier or RP to authenticate the transfer of all session data.

It is important to note that web applications, even those protected by SSL/TLS, can still be vulnerable to a type of session hijacking attack called Cross Site Request Forgery (CSRF). In this type of attack, a malicious website contains a link to the URL of the legitimate RP. The malicious website is generally constructed so that a web browser will automatically send an HTTP request to the RP whenever the browser visits the malicious website. If the Subscriber visits the malicious website while he or she has an open SSL/TLS session with the RP, the request will generally be sent in the same session and with any authentication cookies intact. While the Attacker never gains access to the session secret, the request may be constructed to have side effects, such as sending an email message or authorizing a large transfer of money.

CSRF attacks may be prevented by making sure that neither an Attacker nor a script running on the Attacker's website has sufficient information to construct a valid request authorizing an action (with significant consequences) by the RP. This can be done by inserting random data, supplied by the RP, into any linked URL with side effects and into a hidden field within any form on the RP's website. This mechanism, however, is not effective if the Attacker can run scripts on the RP's website (Cross Site Scripting or XSS). To prevent XSS vulnerabilities, the RP should sanitize inputs from Claimants or Subscribers to make sure they are not executable, or at the very least not malicious, before displaying them as content to the Subscriber's browser.

- *Man-in-the-middle resistance* – Authentication protocols are resistant to a man-in-the-middle attack when both parties (i.e., Claimant and Verifier) are authenticated to the other in a manner that prevents the undetected participation of a third party. There are two levels of resistance:

    1. *Weak man-in-the-middle resistance* – A protocol is said to be weakly resistant to man-in-the-middle attacks if it provides a mechanism for the Claimant to determine whether he or she is interacting with the real Verifier, but still leaves the opportunity for the non-vigilant Claimant to reveal a token authenticator (to an unauthorized party) that can be used to masquerade as the Claimant to the real Verifier. For example, sending a password over server authenticated TLS is weakly resistant to man-in the middle attacks. The browser allows the Claimant to verify the identity of the Verifier; however, if the Claimant is not sufficiently vigilant, the password will be revealed to an unauthorized party who can abuse the information. Weak man-in-the-middle resistance can also be provided by a zero-knowledge password protocol, such as Encrypted Key Exchange (EKE), Simple Password Exponential Key Exchange (SPEKE), or Secure Remote Password Protocol (SRP), which enables the Claimant to authenticate to a Verifier without disclosing the token secret. However, it is possible for the Attacker to trick the Claimant into passing his or her

password into a less secure protocol, thereby revealing the password to the Attacker. Furthermore, if it is unreasonably difficult for the Claimant to verify that the proper protocol is being used, then the overall authentication process does not even provide weak man-in-the-middle resistance (for example, if a zero-knowledge password protocol is implemented by an unsigned java applet displayed on a plaintext HTTP page).

2. *Strong man-in-the-middle resistance*: A protocol is said to be strongly resistant to man-in-the-middle attack if it does not allow the Claimant to reveal, to an Attacker masquerading as the Verifier, information (token secrets, authenticators) that can be used by the latter to masquerade as the true Claimant to the real Verifier. An example of such a protocol is client authenticated TLS, where the browser and the web server authenticate one another using PKI. Even an unwary Claimant cannot easily reveal to an Attacker masquerading as the Verifier any information that can be used by the Attacker to authenticate to the real Verifier. Specialized protocols where the Claimant's token device will only release an authenticator to a preset list of valid Verifiers may also be strongly resistant to man-in-the-middle attacks.

Note that systems can supplement the mitigation strategies listed above by enforcing appropriate security policies. For example, device identity, system health checks, and configuration management can be used to mitigate the risk that the Claimant's system has been compromised.

### 8.2.3. Throttling Mechanisms

When using a token that produces low entropy token Authenticators, it is necessary to implement controls at the Verifier to protect against online guessing attacks. An explicit requirement for such tokens is given in Table 6: the Verifier shall effectively limit online Attackers to 100 failed attempts on a single account in any 30 day period.

The simplest way of implementing a throttling mechanism (which is not the recommended approach) would be to keep a counter of failed attempts that is reset at the beginning of each calendar month, and to lock the account for the rest of the month, when the counter exceeds 50. Aside from the fact that this system would not technically meet the requirement on the first of March in non-leap years, this throttling mechanism has a number of more severe problems. Most notably, it leaves the Verifier open to a very easy denial of service attack (on the first day of the month, an Attacker simply makes 50 failed attempts on each Subscriber account he or she knows about, and the system is unusable for the next 29 days.)

The above simple implementation is also sufficiently limiting that it may suffer from usability problems, where the legitimate Subscriber is penalized for behavior that could reasonably be identified as benign and should not be counted as failed attempts by an Attacker. For example, if the Verifier records a dozen failed authentication attempts followed by a successful attempt from the same IP address over a few minutes to a few hours, it would be reasonable to assume that those attempts did not come from an Attacker.

Additional techniques can be used to prioritize authentication attempts that are likely to come from the Subscriber over those that are more likely to come from an Attacker.

- Requiring the Claimant to complete a Completely Automated Public Turing test to tell Computers and Humans Apart (CAPTCHA) before attempting authentication.

- Requiring the Claimant to wait for a short period of time (anything from 30 seconds to an hour, depending on how close the system is to its maximum allowance for failed attempts) before attempting Authentication following a failed attempt.

- Only accepting authentication requests from a white list of IP addresses at which the Subscriber has been successfully authenticated before.

Since these measures often create user inconvenience, it is best to allow a certain number of failed authentication attempts before employing the above techniques. For example, a system which enforces the 30-day failed attempt limit, by dividing the calendar into 10-day sub-periods and only allowing 25 failed attempts in each sub-period, could allocate failed attempts as follows: in a given 10 day period, the Verifier could allow 2 failed attempts each day regardless of any other considerations, allow an additional 5 failed attempts over the whole period with no additional protections, require CAPTCHAs for the next 5 failed attempts (beyond the 2-per-day quota), and only allow the final 5 attempts to come from a white-listed IP address after the Claimant has completed a CAPTCHA.

Finally, if the Verifier accepts authentication attempts for a large number of Subscribers, it is possible that an Attacker will attempt on online attack on all Subscriber accounts simultaneously, hoping to gain access to one of them, thus circumventing the throttling mechanisms employed on the individual accounts. No specific guideline is given for protecting against such attacks, but Verifiers with a large number of Subscribers should take measures to detect such attacks and either respond to them automatically or alert system administrators to the threat.

### 8.2.4. Phishing and Pharming (Verifier Impersonation): Supplementary Countermeasures

It is important to note that phishing and pharming are attacks that use different techniques to achieve the same goal. Effectively, the Claimant is tricked into believing that he or she is interacting with the Verifier when in actuality, the Verifier is being impersonated by an Attacker attempting to collect token information or other sensitive information.

In a successful phishing attack, the Attacker sends an official looking email to a Subscriber claiming to be a Verifier. The email usually contains a link to a counterfeit Verifier and will ask the Subscriber to click on the link and authenticate to the Verifier[27]. The Subscriber proceeds to authenticate to the counterfeit Verifier and the login information and token authenticator is captured. At this point, the Subscriber is unaware that he or she has been phished, and proceeds with the actions requested by the original email. Once the Subscriber logs off, he or she is unaware that his or her login information has been captured and that potentially sensitive data has been captured.

In a successful pharming attack, the Attacker corrupts either the domain name service (using a technique called DNS poisoning) or the local routing tables (by modifying the host files on a Claimant's computer to point to a bogus DNS server). When the Subscriber attempts to connect to a legitimate Verifier on the Internet, the corrupted DNS tables or routing tables take the Subscriber to a counterfeit Verifier on the Internet. The Subscriber unknowingly reveals token authenticators and other sensitive information to the counterfeit Verifier.

The strongest mechanism for preventing phishing and pharming of authentication secrets, such as token authenticators, is to make sure that some authentication secrets are not directly accessible to the Claimant (as described in Section 8.2.2). However, to help mitigate a wider variety of phishing and pharming attacks, the following techniques may be used:

- *Out of band confirmation of transaction details* – Details (e.g., account number, amount) of sensitive transactions authorized by the Subscriber may be sent by the RP to the Subscriber's out of band token and displayed along with a confirmation code. The confirmation code may either be cryptographically derived from the Subscriber's token secret and the transaction details, or it may be a random value that is sent to the Subscriber's out of band token along with the transaction details. Alternatively, transaction details may be typed in by the Subscriber as manual inputs to a one-time password device. In order to complete the transaction, the Subscriber shall send the correct one-time password or confirmation code to the Verifier or RP.

- *Adding a "Last Login" feature by the Verifier to inform the Subscriber of his or her last login* – If the Subscriber logged in at 8:00am and then logs in at

---

[27] Some phishing attacks may request the Subscriber to provide personally sensitive information so that the Attacker may impersonate the Subscriber outside the scope of E-authentication.

4:00pm but the Last Login feature states that the last login was at 2:00pm, the Subscriber may suspect that he or she has been phished and take appropriate action.

Personalization is the process of customizing a webpage or email for a user to enhance the user experience. For the purpose of this document, personalization schemes can assist the user to determine if he or she is interacting with the correct entity. It is important to note that personalization is at best a low assurance mechanism for mitigating Phishing and Pharming threats, especially when delivered over a communication protocol that is not strongly resistant to man-in-the-middle attacks. However, personalization may provide additional assurance when combined with other techniques.

There are three types of personalization in the context of this guideline:

- *Pre-authentication personalization* – The Verifier displays to the Claimant some personalized indicator (such as an image or user-chosen phrase picked at registration) prior to the latter submitting the token authenticator to the former. This indicator may be established by the Subscriber at the time of registration. When the Claimant views the personalized indicator, the Claimant has an increased sense of assurance that he or she is interacting with the correct Verifier. For example, a Verifier may require the Claimant to submit the username first; in response, the Verifier provides the personalized indicator for the claimed username. If the Claimant recognizes the personalized indicator as his or her own, the Claimant submits his or her token authenticator to the Verifier. Pre-authentication personalization does not eliminate Phishing attacks, but requires the Attacker to use a more complex technique to succeed in a Phishing attack.

- *Post-authentication personalization* – The Verifier displays a personalized indicator to the Subscriber after successful authentication of the latter. The personalized indicator provides assurance to the Subscriber that he or she has in fact logged in to the correct site. This indicator may be established by the Subscriber at the time of registration. For example, after a Subscriber authenticates to the Verifier, the Verifier provides a personalized indicator (such as a picture, a phrase, or a greeting) that the Subscriber can readily recognize as his or her own. If the personalized indicator is not shown, or is not recognized by the Subscriber, the Subscriber suspects that he or she has been phished and takes appropriate action. Post-authentication personalization does not protect any secrets used by the Subscriber in the initial authentication process. Nonetheless, if some or all of these secrets are protected by hardware or software that runs a protocol with strong man-in-the-middle resistance, then the personalization will assist the Subscriber in recognizing that he or she is interacting with a bogus site and refraining from revealing any further sensitive information. If personalization appears before the Subscriber is prompted for a password, but after the Verifier strongly authenticates the

Subscriber's local system, then the Subscriber's password may also be protected from phishing.

- *Personalization of email sent to the Subscriber by a valid Verifier* – This type of personalization is employed to help the Subscriber differentiate between email from a valid Verifier, and email from a Phisher. For example, an email from a Verifier may contain a picture which the Subscriber selected in the registration process. This type of personalization forces the Phisher to use a fairly difficult attack and in effect forces the Phisher to either use a targeted attack against each Subscriber or hope that the Subscriber will not notice the incorrect or missing personalization identifier.

It is important to note that using a Subscriber's name (first or last) as the only method of personalization is a relatively weak method to thwart a phishing attack since it is fairly easy for an Attacker to gain this type of information and display it in an email or display it after logging into a site. Information of a non-public nature is a better candidate for use during personalization.

### 8.3. Authentication Process Assurance Levels

The stipulations for authentication process assurance levels are described in the following sections.

#### 8.3.1. Threat Resistance per Assurance Level

Authentication process assurance levels can be defined in terms of required threat resistance. Table 11 lists the threat resistance requirements per assurance level:

Table 11 – Required Authentication Protocol Threat Resistance per Assurance Level

| Authentication Process Attacks/Threats | Threat Resistance Requirements | | | |
|---|---|---|---|---|
| | Level 1 | Level 2 | Level 3 | Level 4 |
| Online guessing | Yes | Yes | Yes | Yes |
| Replay | Yes | Yes | Yes | Yes |
| Session hijacking | No | Yes | Yes | Yes |
| Eavesdropping | No | Yes | Yes | Yes |
| Phishing/pharming(verifier impersonation) | No | No | Yes[28] | Yes |
| Man in the middle | No | Weak | Weak | Strong |
| Denial of service/flooding[29] | No | No | No | No |

---

[28] Long term authentication secrets shall be protected at this level. Short term secrets may or may not be protected.
[29] Although there are techniques used to resist flood attacks, no protocol has comprehensive resistance to stop flooding.

### 8.3.2. Requirements per Assurance Level

This section states the requirements levied on the authentication process to achieve the required threat resistance at each assurance level. At Levels 2 and above, the authentication process shall provide sufficient information to the Verifier to uniquely identify the appropriate registration information that was (i) provided by the Subscriber at the time of registration, and (ii) verified by the RA in the issuance of the token and credential. It is important to note that the requirements listed below will not protect the authentication process if malicious code is introduced on the Claimant's machine or at the Verifier.

#### 8.3.2.1. Level 1

Although there is no identity proofing requirement at this level, the authentication mechanism provides some assurance that the same Claimant who participated in previous transactions is accessing the protected transaction or data. It allows a wide range of available authentication technologies to be employed and permits the use of any of the token methods of Levels 2, 3 or 4. Successful authentication requires that the Claimant prove, through a secure authentication protocol, that he or she possesses and controls the token.

Plaintext passwords or secrets shall not be transmitted across a network at Level 1. However this level does not require cryptographic methods that block offline analysis by eavesdroppers. For example, password challenge-response protocols that combine a password with a challenge to generate an authentication reply satisfy this requirement although an eavesdropper who intercepts the challenge and reply may be able to conduct a successful off-line dictionary or password exhaustion attack and recover the password. Since an eavesdropper who intercepts such a protocol exchange will often be able to find the password with a straightforward dictionary attack, and this vulnerability is independent of the strength of the operations, there is no requirement at this level to use Approved cryptographic techniques. At Level 1, long-term shared authentication secrets may be revealed to Verifiers.

A wide variety of technologies should be able to meet the requirements of Level 1. For example, a Verifier might obtain a Subscriber password from a CSP and authenticate the Claimant by use of a challenge-response protocol. A password sent through a TLS protocol session is another example. Other common protocols that meet Level 1 requirements include APOP [RFC 1939], S/KEY [RFC 1760], and password-based versions of Kerberos [KERB].

#### 8.3.2.2. Level 2

Level 2 allows a wide range of available authentication technologies to be employed and permits the use of any of the token methods of Levels 2, 3 and 4. Successful authentication requires that the Claimant shall prove, through a secure authentication protocol, that he or she controls the token. Session hijacking (when required based on the FIPS 199 security category of the systems as described below), replay, and online guessing attacks shall be resisted. Approved cryptography is required to resist eavesdropping to capture authentication data. Protocols used at Level 2 and above shall be at least weakly man-in-the-middle resistant, as described in the threat mitigation strategies subsection.

Session data transmitted between the Claimant and the RP following a successful Level 2 authentication shall be protected as described in the NIST FISMA guidelines. Specifically, all session data exchanged between information systems that are categorized as FIPS 199 "Moderate" or "High" for confidentiality and integrity, shall be protected in accordance with NIST SP 800-53 Control SC-8 (which requires transmission confidentiality) and SC-9 (which requires transmission integrity).

A wide variety of technologies can meet the requirements of Level 2. For example, a Verifier might authenticate a Claimant who provides a password through a secure (encrypted) TLS protocol session (tunneling).

### 8.3.2.3. Level 3

Level 3 provides multi-factor remote network authentication. At least two authentication factors are required. Level 3 authentication is based on proof of possession of the allowed types of tokens through a cryptographic protocol. Level 3 also permits any of the token methods of Level 4. Refer to Section 6 for requirements for single tokens and token combinations that can achieve Level 3 authentication assurance. Additionally, At Level 3, strong cryptographic mechanisms shall be used to protect token secret(s) and authenticator(s). Long-term shared authentication secrets, if used, shall never be revealed to any party except the Claimant and CSP; however, session (temporary) shared secrets may be provided to Verifiers by the CSP, possibly via the Claimant. Approved cryptographic techniques shall be used for all operations including the transfer of session data.

Level 3 assurance may be satisfied by client authenticated TLS (implemented in all modern browsers), with Claimants who have public key certificates. Other protocols with similar properties may also be used.

Level 3 authentication assurance may also be met by tunneling the output of a MF OTP Token, or the output of a SF OTP Token in combination with a Level 2 personal password, through a TLS session.

### 8.3.2.4. Level 4

Level 4 is intended to provide the highest practical remote network authentication assurance. Refer to Section 6 for single tokens and token combinations that are allowed to be used to achieve Level 4 authentication assurance.

Level 4 requires strong cryptographic authentication of all parties, and all sensitive data transfers between the parties. Either public key or symmetric key technology may be used. The token secret shall be protected from compromise through the malicious code threat as described in Section 0 above. Long-term shared authentication secrets, if used, shall never be revealed to any party except the Claimant and CSP; however session (temporary) shared secrets may be provided to Verifiers or RPs by the CSP. Strong, Approved cryptographic techniques shall be used for all operations including the transfer of session data. All sensitive data transfers shall be cryptographically authenticated using keys that are derived from the authentication process in such a way that MitM attacks are strongly resisted.

Level 4 assurance may be satisfied by client authenticated TLS (implemented in all modern browsers), with Claimants who have public key MF Hardware Cryptographic Tokens. Other protocols with similar properties can also be used.

It should be noted that, in multi-token schemes, the token used to provide strong man-in-the-middle resistance need not be a hardware token. For example, if a software cryptographic token is used to open a client-authenticated TLS session, and the output of a multifactor OTP device is sent by the claimant in that session, then the resultant protocol will still provide Level 4 assurance.

# 9. Assertions

## 9.1. Overview

Assertions are statements from a Verifier to an RP that contain information about a Subscriber. Assertions are used when the RP and the Verifier are not collocated (i.e., they are connected through a shared network). The RP uses the information in the assertion to identify the Claimant and make authorization decisions about his or her access to resources controlled by the RP. An assertion may include identification and authentication statements regarding the Subscriber, and may additionally include attribute statements that further characterize the Subscriber and support the authorization decision at the RP.

Assertion-based authentication of the Claimant serves several important goals. It supports the process of Single-Sign-On for Claimants, allowing them to authenticate once to a Verifier and subsequently obtain services from multiple RPs without being aware of further authentication. Assertion mechanisms also support the implementation of a federated identity for a Subscriber, allowing the linkage of multiple identities/accounts held by the Subscriber with different RPs through the use of a common "federated" identifier. In this context, a federation is a group of entities (RPs, Verifiers and CSPs) that are bound together through common agreed-upon business practices, policies, trust mechanisms, profiles and protocols. Finally, assertion mechanisms can also facilitate authentication schemes that are based on the attributes or characteristics of the Claimant in lieu of (or in addition to) the identity of the Claimant. Attributes are often used in determining access privileges for Attributes Based Access Control (ABAC) or Role Based Access Control (RBAC).

It is important to note that assertion schemes are fairly complex multiparty protocols, and therefore have fairly subtle security requirements which shall be satisfied. When evaluating a particular assertion scheme, it may be instructive to break it down into its component interactions. Generally speaking, interactions between the Claimant/Subscriber and the Verifier and between the Claimant/Subscriber and RP are similar to the authentication mechanisms presented in Section 8, while interactions between the Verifier and RP are similar to the *token and credential verification services* presented in Section 7. Many of the requirements presented in this section will, therefore, be similar to corresponding requirements in those two sections.

There are two basic models for assertion-based authentication. After successful authentication with the Verifier, the Subscriber is issued an assertion or an assertion reference, which the Subscriber uses to authenticate to the RP.

- *The Direct Model* – In the direct model, the Claimant uses his or her e-authentication token to authenticate to the Verifier. Following successful authentication of the Claimant, the Verifier creates an assertion, and sends it to the Subscriber to be forwarded to the RP. The assertion is used by the

Claimant/Subscriber to authenticate to the RP. (This is usually handled automatically by the Subscriber's browser.) Figure 4 illustrates this model.

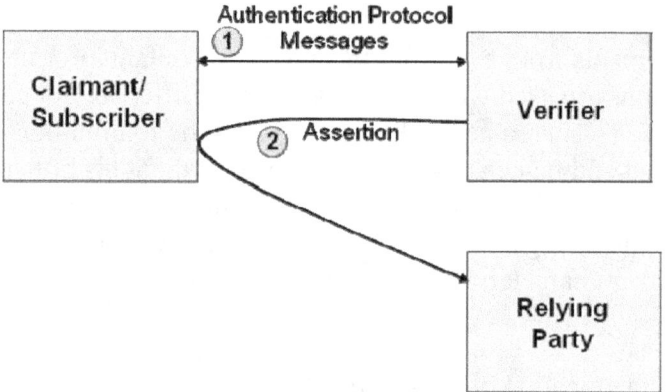

Figure 4 - *Direct Assertion Model*

- *The Indirect Model* – In the indirect model, the Claimant uses his or her token to authenticate to the Verifier. Following successful authentication, the Verifier creates an assertion as well as an assertion reference (which identifies the Verifier and includes a pointer to the full assertion held by the Verifier). The assertion reference is sent to the Subscriber to be forwarded to the RP. In this model, the assertion reference is used by the Claimant/Subscriber to authenticate to the RP. The RP then uses the assertion reference to explicitly request the assertion from the Verifier. Figure 5 illustrates this model.

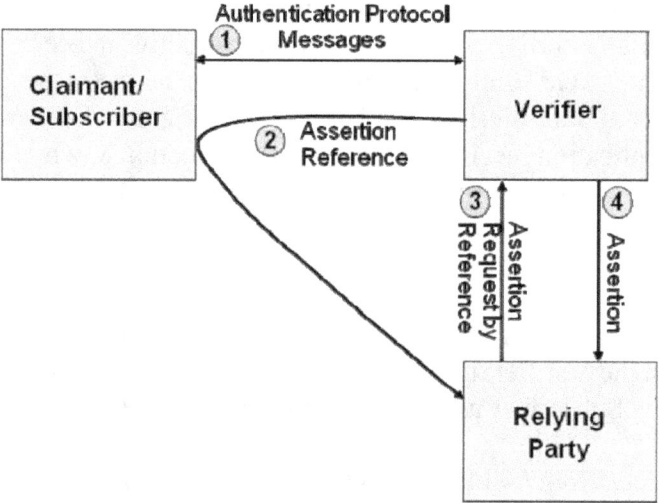

Figure 5 - *Indirect Assertion Model*

As mentioned earlier, an assertion contains a set of claims or statements about an authenticated Subscriber. Based on the statements contained within it, an authentication assertion will fall into one of two categories (and either category can be used in both direct and indirect models):

- *Holder-of-Key Assertions* – A holder-of-key assertion contains a reference to a symmetric key or a public key (corresponding to a private key) possessed by the Subscriber. The RP may require the Subscriber to prove possession of the secret that is referenced in the assertion. In proving possession of the Subscriber's secret, the Subscriber also proves with a certain degree of assurance that he or she is the rightful owner of the assertion. It is therefore difficult for an Attacker to use a holder-of-key assertion issued to a Subscriber, since the former cannot prove possession of the secret referenced within the assertion.

- *Bearer Assertions* – A bearer assertion does not provide a mechanism for the Claimant to prove that he or she is the rightful owner of the assertion. The RP has to assume that the assertion was issued to the Subscriber who presents the assertion or the corresponding assertion reference to the RP. If a bearer assertion (in the direct model) or assertion reference (in the indirect model) belonging to a Subscriber is captured, copied, or manufactured by an Attacker, the latter can impersonate the rightful Subscriber to obtain services from the RP. Bearer assertions can be made secure only if some part of the assertion or assertion reference, sent to the Subscriber by the Verifier, is unpredictable to an Attacker and can reliably be kept secret.

There are cases in which the RP should be anonymous to the Verifier for the purpose of privacy. The direct model is more suitable for the "anonymous RP" scenario since there is no requirement for the RP to authenticate to the Verifier as in the indirect model. However, it is possible to devise authentication schemes (e.g., using key hierarchies within a group or federation) that allow the use of the indirect model to support the "anonymous RP" scenario.

There are other cases where privacy concerns require that the Claimant's identity/account at the Verifier and RP not be linked through use of a common identifier/account name. In such scenarios, pseudonymous identifiers are used within the assertions generated by the Verifier for the RP.

It should be noted that the two models described above are abstractions. There may be other interactions between the three players preceding or interspersed with the interactions described in the model. For example, the Claimant may initiate a connection with an RP of his or her choice, at which point, the latter would redirect the Claimant to an appropriate Verifier to be authenticated using the direct model, resulting in an assertion being sent to the RP. Alternately, the Claimant may first authenticate to a Verifier of his or her choice and then select one or more RPs to obtain further services.

The direct model is used to generate assertions for each of these RPs. Parallel scenarios may be constructed for the indirect model as well.

There is one other basic assertion model.

- *The Proxy Model* – In the proxy model, the Claimant uses his or her e-authentication token to authenticate to the Verifier. Following successful authentication of the Claimant, the Verifier creates an assertion and includes it when interacting directly with the RP, acting as an intermediary between the Claimant and the RP. Figure 6 illustrates this model.

Figure 6 – *Proxy Model*

The RP grants or denies the request based, at least in part, on the authentication assertion made by the Verifier. There are several common reasons for such proxies:
- o Portals that provide users access to multiple RPs that require user authentication
- o Web caching mechanisms that are required to satisfy the RP's access control policies, especially when client-authenticated TLS with the Claimant is required
- o Network monitoring and/or filtering mechanisms that terminate TLS in order to inspect and manipulate the traffic

It is good practice to protect communications between the Verifier and the RP. Current commercial implementations tend to do this by having the proxy use client-authenticated TLS with the Verifier and pass the authentication assertion in the HTTP header.

Note that the Verifier may have access to information that may be useful to the RP in enforcing security policies, such as device identity, location, system health checks, and configuration management. If so, it may be a good idea to pass this information along to the RP.

Three types of assertion technologies will be discussed within this section: Web browser cookies, SAML (Security Assertion Markup Language) assertions, and Kerberos tickets.

Other assertion technologies may be used in an e-authentication environment as long as they meet the requirements set forth in Section 9.3 below for the targeted assurance level.

### 9.1.1. Cookies

One type of assertion widely in use is Web cookie technology. Cookies are text files used by a browser to store information provided by a particular web site. The contents of the cookie are sent back to the web site each time the browser requests a page from the same web site. The web site uses the contents of the cookie to identify the user and prepare customized Web pages for that user, or to authorize the user for certain transactions.

Cookies have two mandatory parameters:

- *Name* – This parameter states the name of the cookie.
- *Value* – This parameter holds information that a cookie is storing. For example, the value parameter could hold a user ID or session ID.

Cookies also have four optional parameters:

- *Expiration date* – This parameter determines how long the cookie stays valid.
- *Path* – This parameter sets the path over which the cookie is valid.
- *Domain* – This parameter determines the domain in which the cookie is valid.
- *Secure* – This parameter indicates the cookie requires that a secure connection exist for the cookie to be used.

There are two types of cookies:

- *Session cookies* – A cookie that is erased when the user closes the web browser. The session cookie is stored in temporary memory and is not retained after the browser is closed.
- *Persistent cookies* – A cookie that is stored on a user's hard drive until it expires (persistent cookies are set with expiration dates) or until the user deletes the cookie.

Cookies are effective as assertions for Internet single-sign-on where the RP and Verifier are part of the same Internet domain, and when the cookie contains authentication status for that domain. They are not usable in scenarios where the RP and the Verifier are part of disparate domains.

Cookies are also often used by the Claimant to re-authenticate to a server. This may be considered to be a use of assertion technology. In this case, the server acts as a Verifier when it sets the cookie in the Subscriber's browser, and as an RP when it requests the cookie from a Claimant who wishes to re-authenticate to it. Often, the cookie contains a random number, and the assertion data that it represents does not leave the server. Note that, if the cookie is used as an assertion reference in this way, no assertion needs to be sent on an open network, and therefore, confidentiality and integrity requirements for assertion data at Level 2 and below may be satisfied by access controls rather than by cryptographic methods. (The cookie itself, however, does need to be protected.) This is in line with the credential storage requirement presented in Section 7.

### 9.1.2. Security Assertion Markup Language (SAML) Assertions

SAML is an XML-based framework for creating and exchanging authentication and attribute information between trusted entities over the Internet. As of this writing, the latest specification for [SAML] is SAML v2.0, issued 15 March 2005.

The building blocks of SAML include the Assertions XML schema which define the structure of the assertion; the SAML Protocols which are used to request assertions and artifacts (that is, the assertion reference mentioned in Section 9.1); and the Bindings that define the underlying communication protocols (such as HTTP or SOAP) and that can be used to transport the SAML assertions. The three components above define a SAML profile that corresponds to a particular use case such as "Web Browser SSO".

SAML Assertions are encoded in an XML schema and can carry up to three types of statements:

- *Authentication statements* – Include information about the assertion issuer, the authenticated subject, validity period, and other authentication information. For example, an Authentication Assertion would state the subject "John" was authenticated using a password at 10:32pm on 06-06-2004.

- *Attribute statements* – Contain specific additional characteristics related to the Subscriber. For example, subject "John" is associated with attribute "Role" with value "Manager".

- *Authorization statements* – Identify the resources the Subscriber has permission to access. These resources may include specific devices, files, and information on specific web servers. For example, subject "John" for action "Read" on "Webserver1002" given evidence "Role".

Authorization statements are beyond the scope of this document and will not be discussed.

### 9.1.3. Kerberos Tickets

The Kerberos Network Authentication Service [RFC 4120] was designed to provide strong authentication for client/server applications using symmetric-key cryptography. Extensions to Kerberos can support the use of public key cryptography for selected steps of the protocol. Kerberos also supports confidentiality and integrity protection of session data between the Subscriber and the RP.

Kerberos supports authentication of a Claimant over an untrusted, shared network using two or more Verifiers. The Claimant implicitly authenticates to the Verifier by demonstrating the ability to decrypt a random session key encrypted for the Subscriber by the Verifier. (Some Kerberos variants also require the Subscriber to explicitly authenticate to the Verifier, but this is not universal.) In addition to the encrypted session key, the Verifier also generates another encrypted object called a Kerberos ticket. The ticket contains the same session key, the identity of the Subscriber to whom the session key was issued, and an expiration time after which the session key is no longer valid. The ticket is confidentiality and integrity protected by a pre-established that is key shared between the Verifier and the RP.

To authenticate using the session key, the Claimant sends the ticket to the RP along with encrypted data that proves that the Claimant possesses the session key embedded within the Kerberos ticket. Session keys are either used to generate new tickets, or to encrypt and authenticate communications between the Subscriber and the RP.

To begin the process, the Claimant sends an authentication request to the Authentication Server (AS). The AS encrypts a session key for the Subscriber using the Subscriber's long term credential. The long term credential may either be a secret key shared between the AS and the Subscriber, or in the PKINIT variant of Kerberos, a public key certificate. It should be noted that most variants of Kerberos based on a shared secret key between the Subscriber and Verifier derive this key from a user generated password. As such, they are vulnerable to offline dictionary attack by a passive eavesdropper.

In addition to delivering the session key to the subscriber, the AS also issues a ticket using a key it shares with the Ticket Granting Server (TGS). This ticket is referred to as a Ticket Granting Ticket (TGT), since the verifier uses the session key in the TGT to issue tickets rather than to explicitly authenticate the Claimant. The TGS uses the session key in the TGT to encrypt a new session key for the Subscriber and uses a key it shares with the RP to generate a ticket corresponding to the new session key. The subscriber decrypts the session key and uses the ticket and the new session key together to authenticate to the RP.

### 9.2. Assertion Threats

In this section, it is assumed that the two endpoints of the assertion transmission (namely, the Verifier and the RP) are uncompromised. However, the Claimant is not assumed to be entirely trustworthy as the Claimant may have an interest in modifying or replacing an assertion to obtain a greater level of access to a resource/service provided by the RP. Other Attackers are assumed to lurk within the shared transmission medium (e.g., Internet) and may be interested in obtaining or modifying assertions and assertion references to impersonate a Subscriber or access unauthorized data or services.

Furthermore, it is possible that two or more entities may be colluding to attack another party. An Attacker may attempt to subvert assertion protocols by directly compromising the integrity or confidentiality of the assertion data. For the purpose of this type of threat, authorized parties who attempt to exceed their privileges may be considered Attackers.

- *Assertion manufacture/modification* – An Attacker may generate a bogus assertion or modify the assertion content (such as the authentication or attribute statements) of an existing assertion, causing the RP to grant inappropriate access to the Subscriber. For example, an Attacker may modify the assertion to extend the validity period; a Subscriber may modify the assertion to have access to information that they should not be able to view.

- *Assertion disclosure* – Assertions may contain authentication and attribute statements that include sensitive Subscriber information. Disclosure of the assertion contents can make the Subscriber vulnerable to other types of attacks.

- *Assertion repudiation by the Verifier* – An assertion may be repudiated by a Verifier if the proper mechanisms are not in place. For example, if a Verifier does not digitally sign an assertion, the Verifier can claim that it was not generated through the services of the Verifier.

- *Assertion repudiation by the Subscriber* – Since it is possible for a compromised or malicious subscriber to issue assertions to the wrong party, a subscriber can repudiate any transaction with the RP that was authenticated using only a bearer assertion.

- *Assertion redirect*: An Attacker uses the assertion generated for one RP to obtain access to a second RP.

- *Assertion reuse* – An Attacker attempts to use an assertion that has already been used once with the intended RP.

In addition to reliable and confidential transmission of assertion data from the Verifier to the RP, assertion protocols have a further goal: in order for the Subscriber to be recognized by the RP, he or she shall be issued some secret information, the knowledge of which distinguishes the Subscriber from Attackers who wish to impersonate the Subscriber. In the case of holder-of-key assertions, this secret is generally the Subscriber's long term token secret, which would already have been established with the CSP prior to the initiation of the assertion protocol.[30]

In other cases, however, the Verifier will generate a temporary secret and transmit it to the authenticated Subscriber for this purpose. Since, when this secret is used to authenticate to the RP, it generally replaces the token authenticator in the type of

---

[30] The role of the Verifier in such protocols is not necessarily to issue new secrets. Rather, in a holder-of-key-assertion, the Verifier communicates the information in the Subscriber's credential (as well as any supplementary information from the CSP such as revocation data) to the RP. The Verifier also vouches that the holder-of-key assertion represents current information from a trusted source (the CSP.)

protocols described in Section 8, this temporary secret will be referred to here as a secondary authenticator. Secondary authenticators include assertions in the direct model, session keys in Kerberos, assertion references in the indirect model, and cookies used for authentication. The threats to the secondary authenticator are as follows:

- *Secondary authenticator manufacture* – An Attacker may attempt to generate a valid secondary authenticator and use it to impersonate a Subscriber.

- *Secondary authenticator capture* – The Attacker may use a session hijacking attack to capture the secondary authenticator when the Verifier transmits it to the Subscriber after the primary authentication step, or the Attacker may use a man-in-the-middle attack to obtain the secondary authenticator as it is being used by the Subscriber to authenticate to the RP. If, as in the indirect model, the RP needs to send the secondary authenticator back to the Verifier in order to check its validity or obtain the corresponding assertion data, an Attacker may similarly subvert the communication protocol between the Verifier and the RP to capture a secondary authenticator. In any of the above scenarios, the secondary authenticator can be used to impersonate the Subscriber.

Finally, in order for the Subscriber's authentication to the RP to be useful, the binding between the secret used to authenticate to the RP and the assertion data referring to the Subscriber shall be strong.

- *Assertion substitution* – A subscriber may attempt to impersonate a more privileged subscriber by subverting the communication channel between the Verifier and RP, for example by reordering the messages, to convince the RP that his or her secondary authenticator corresponds to assertion data sent on behalf of the more privileged subscriber. This is primarily a threat to the indirect model, since in the direct model, assertion data is directly encoded in the secondary authenticator.

### 9.2.1. Threat Mitigation Strategies

Mitigation techniques are described below for each of the threats described in the last subsection.

Logically speaking, an assertion is issued by a Verifier and consumed by an RP – these are the two end points of the session that needs to be secured to protect the assertion. In the direct model, the session in which the assertion is passed traverses the Subscriber. Furthermore, in the current web environment, the assertion may pass through two separate secure sessions (one between the Verifier and the Subscriber, and the other between the Subscriber and the RP), with a break in session security on the Subscriber's browser. This is reflected in the mitigation strategies described below. In the indirect model, the assertion flows directly from the Verifier to the RP; this protocol session needs to be protected. All of the threat mitigation strategies in Section 8 apply to the protocols used to request, retrieve and submit assertions and assertion references.

- *Assertion manufacture/modification*: To mitigate this threat, one of the following mechanisms may be used:

    1. The assertion may be digitally signed by the Verifier. The RP should check the digital signature to verify that it was issued by a legitimate Verifier.

    2. The assertion may be sent over a protected session such as TLS/SSL. In order to protect the integrity of assertions from malicious attack, the Verifier shall be authenticated.

- *Assertion disclosure* – To mitigate this threat, one of the following mechanisms may be implemented:

    1. The assertion may be sent over a protected session to an authenticated RP. Note that, in order to protect assertions against both disclosure and manufacture/modification using a protected session, both the RP and the Verifier need to be authenticated.

    2. If assertions are signed by the Verifier, they may be encrypted for a specific RP with no additional integrity protection. It should be noted that any protocol that requires a series of messages between two parties to be signed by their source and encrypted for their recipient provides all the same guarantees as a mutually authenticated protected session, and may therefore be considered equivalent. The general requirement for protecting against both assertion disclosure and assertion manufacture/modification may therefore be described as a mutually authenticated protected session or equivalent between Verifier and RP.

- *Assertion repudiation by the Verifier* – To mitigate this threat, the assertion may be digitally signed by the Verifier using a key that supports non-repudiation. The RP should check the digital signature to verify that it was issued by a legitimate Verifier.

- *Assertion repudiation by the Subscriber* – To mitigate this threat, the Verifier may issue holder of key, rather than bearer assertions. The Subscriber can then prove possession of the asserted key to the RP. If the asserted key matches the subscriber's long term credential (as provided by the CSP) it will be clear to all parties involved that it was the Subscriber who authenticated to the RP rather than a compromised Verifier impersonating the Subscriber.

- *Assertion redirect* – To mitigate this threat, the assertion may include the identity of the RP for whom it was generated. The RP verifies that incoming assertions include its identity as the recipient of the assertion.

- *Assertion reuse* – To mitigate this threat, the following mechanisms may be used:
    1. The assertion includes a timestamp and has a short lifetime of validity. The RP checks the timestamp and lifetime values to ensure that the assertion is currently valid. The lifetime value may either be in the assertion or set by the RP.
    2. The RP keeps track of assertions that were consumed within a (configurable) time window to ensure that an assertion cannot be used more than once within that time window.

- *Secondary authenticator manufacture* – To mitigate this threat, one of the following mechanisms may be implemented:
    1. The secondary authenticator may contain sufficient entropy that an Attacker without direct access to the Verifier's random number generator cannot guess the value of a valid secondary authenticator.
    2. The secondary authenticator may contain timely assertion data that is signed by the Verifier or integrity protected using a key shared between the Verifier and the RP.
    3. The Subscriber may authenticate to the RP directly using his or her long term token and avoid the need for a secondary authenticator altogether.

- *Secondary authenticator capture* – To mitigate this threat, adequate protections shall be in place throughout the lifetime of any secondary authenticators used in the assertion protocol.
    1. In order to protect the secondary authenticator while it is in transit between the Verifier and the Subscriber, the secondary authenticator shall be sent via a protected session established during the primary authentication of the Subscriber using his or her token. This requirement is the same as the requirement in Section 8, regarding the Authentication Process, to protect sensitive data (in this case the secondary authenticator) from session hijacking attacks.
    2. In order to protect the secondary authenticator from capture as it is submitted to the RP, the secondary authenticator shall be used in an authentication protocol which protects against eavesdropping and man-in-the-middle attacks as described in Section 8.
    3. In order to protect the secondary authenticator after it has been used, it shall never be transmitted on an unprotected session or to an unauthenticated party while it is still valid. The secondary authenticator may be sent in the clear only if the sending party has strong assurances that the secondary authenticator will not subsequently be accepted by any other RP. This is possible if the secondary authenticator is specific to a

single RP, and if that RP will not accept secondary authenticators with the same value until the maximum lifespan of the corresponding assertion has passed.

- *Assertion substitution* – To mitigate this threat, one of the following mechanisms may be implemented:
  1. Responses to assertion requests, signed or integrity protected by the Verifier, may contain the value of the assertion reference used in the request or some other nonce that was cryptographically bound to the request by the RP.

  2. Responses to assertion requests may be bound to the corresponding requests by message order, as in HTTP, provided that assertions and requests are protected by a protocol such as TLS that can detect and disallow malicious reordering of packets.

## 9.3. Assertion Assurance Levels

The stipulations for assertion assurance levels are described in the next sections.

### 9.3.1. Threat Resistance per Assurance Level

Table 12 lists the requirements for assertions (both in the direct and indirect models) and assertion references (in the indirect model) at each assurance level in terms of resistance to the threats listed above.

Table 12 – Threat Resistance per Assurance Level

| Threat | Level 1 | Level 2 | Level 3 | Level 4 |
|---|---|---|---|---|
| Assertion manufacture/modification | Yes | Yes | Yes | Yes |
| Assertion disclosure | No | Yes | Yes | Yes |
| Assertion repudiation by Verifier | No | No | Yes[31] | Yes[31] |
| Assertion repudiation by Subscriber | No | No | No | Yes[31] |
| Assertion redirect | No | Yes | Yes | Yes |
| Assertion reuse | Yes | Yes | Yes | Yes |
| Secondary authenticator manufacture | Yes | Yes | Yes | Yes |
| Secondary authenticator capture | No | Yes | Yes | Yes |
| Assertion substitution | No | Yes | Yes | Yes |

---

[31] Except for Kerberos.

### 9.3.2. Requirements per Assurance Level

The following sections summarize the requirements for assertions at each assurance level.

All assertions recognized within this guideline shall indicate the assurance level of the initial authentication of the Claimant to the Verifier. The assurance level indication within the assertion may be implicit (e.g., through the identity of the Verifier implicitly indicating the resulting assurance level) or explicit (e.g., through an explicit field within the assertion).

#### 9.3.2.1. Level 1

At Level 1, it must be impractical for an Attacker to manufacture an assertion or assertion reference that can be used to impersonate the Subscriber. If the direct model is used, the assertion which is used shall be signed by the Verifier or integrity protected using a secret key shared by the Verifier and RP, and if the indirect model is used, the assertion reference which is used shall have a minimum of 64 bits of entropy. Bearer assertions shall be specific to a single transaction.[32] Also, if assertion references are used, they shall be freshly generated whenever a new assertion is created by the Verifier. In other words, bearer assertions and assertion references are generated for one-time use.

Furthermore, in order to protect assertions against modification in the indirect model, all assertions sent from the Verifier to the RP shall either be signed by the Verifier, or transmitted from an authenticated Verifier via a protected session. In either case, a strong mechanism must be in place which allows the RP to establish a binding between the assertion reference and its corresponding assertion, based on integrity protected (or signed) communications with the authenticated Verifier.

To lessen the impact of captured assertions and assertion references, assertions that are consumed by an RP which is not part of the same Internet domain as the Verifier shall expire if they are not used within 5 minutes of their creation. Assertions intended for use within a single Internet domain, including assertions contained in or referenced by cookies, however, may last as long as 12 hours without being used.

#### 9.3.2.2. Level 2

If the underlying credential specifies that the subscriber name is a pseudonym, this information must be conveyed in the assertion. Level 2 assertions shall be protected against manufacture/modification, capture, redirect and reuse. Assertion references shall be protected against manufacture, capture and reuse. Each assertion shall be targeted for a

---

[32] For example, implementation of SSO requires a separate assertion each time a new session is started with a participating RP.

single RP and the RP shall validate that it is the intended recipient of the incoming assertion.

All stipulations from Level 1 apply. Additionally, assertions, assertion references and any session cookies used by the Verifier or RP for authentication purposes, shall be transmitted to the Subscriber through a protected session which is linked to the primary authentication process in such a way that session hijacking attacks are resisted (see Section 8.2.2 for methods which may be used to protect against session hijacking attacks). Assertions, assertion references and session cookies shall not be subsequently transmitted over an unprotected session or to an unauthenticated party while they remain valid. (To this end, any session cookies used for authentication purposes shall be flagged as secure, and redirects used to forward secondary authenticators from the Subscriber to the RP shall specify a secure protocol such as HTTPS.)

To protect assertions against manufacture, modification, and disclosure, assertions which are sent from the Verifier to the RP, whether directly or through the Subscriber's device, shall either be sent via a mutually authenticated protected session between the Verifier and RP, or equivalently shall be signed by the Verifier and encrypted for the RP.

All assertion protocols used at Level 2 and above require the use of Approved cryptographic techniques. As such, the use of Kerberos keys derived from user generated passwords is not permitted at Level 2 or above.

### 9.3.2.3. Level 3

At Level 3, in addition to Level 2 requirements, assertions shall be protected against repudiation by the Verifier; all assertions used at Level 3 shall be signed. Level 3 assertions shall specify verified names and not pseudonyms.

Kerberos uses symmetric key mechanisms to protect key management and session data, and it does not protect against assertion repudiation. However, based on the high degree of vetting conducted on the Kerberos protocol and its wide deployment, Kerberos tickets are acceptable for use as assertions at Level 3 as long as:

- All Verifiers (Kerberos Authentication Servers and Ticket Granting Servers) are under the control of a single management authority that ensures the correct operation of the Kerberos protocol;

- The Subscriber authenticates to the Verifier using a Level 3 token;

- All Level 3 requirements unrelated to non-repudiation are satisfied.

Also, at Level 3, single-domain assertions (e.g., Web browser cookies) shall expire if they are not used within 30 minutes. Cross-domain assertions shall expire if not used within 5 minutes.

However, in order to deliver the effect of single sign on, the Verifier may re-authenticate the Subscriber prior to delivering assertions to new RPs, using a combination of long term and short term single domain assertions provided that the following assurances are met:

- The Subscriber has successfully authenticated to the Verifier within the last 12 hours.

- The Subscriber can demonstrate that he or she was the party that authenticated to the Verifier. This could be demonstrated, for example, by the presence of a cookie set by the Verifier in the Subscriber's browser.

- The Verifier can reliably determine whether the Subscriber has been in active communication with an RP since the last assertion was delivered by the Verifier. This means that the Verifier needs evidence that the Subscriber is actively using the services of the RP and has not been idle for more than 30 minutes. An authenticated assertion by the RP to this effect is considered sufficient evidence for this purpose.

### 9.3.2.4. Level 4

At Level 4, bearer assertions (including cookies) shall not be used to establish the identity of the Claimant to the RP. Assertions made by the Verifier may however be used to bind keys or other attributes to an identity. Holder-of-key assertions may be used, provided that all three requirements below are met:

- The Claimant authenticates to the Verifier using a Level 4 token (as described in Section 6) in a Level 4 authentication protocol (as described in Section 8).

- The Verifier generates a holder-of-key assertion that references a key that is part of the Level 4 token (used to authenticate to the Verifier) or linked to it through a chain of trust, and;

- The RP verifies that the Subscriber possesses the key that is referenced in the holder-of-key assertion using a Level 4 protocol (where the RP plays the role attributed to the Verifier by Section 8).

The RP should maintain records of the assertions it receives, so that if a suspicious transaction occurs at the RP, the key asserted by the Verifier may be compared to the value registered with the CSP. This record keeping allows the RP to detect any attempt by the Verifier to impersonate the Subscriber using fraudulent assertions and may also be useful for preventing the Subscriber from repudiating various aspects of the authentication process.

Kerberos uses symmetric key mechanisms to protect key management and session data, and it does not protect against assertion repudiation by the Subscriber or the Verifier. However, based on the high degree of vetting conducted on the Kerberos protocol and its wide deployment, Kerberos tickets are acceptable for use as assertions at Level 4 as long as:

- All Verifiers (Kerberos Authentication Servers and Ticket Granting Servers) are under the control of a single management authority that ensures the correct operation of the Kerberos protocol;

- The Subscriber authenticates to the Verifier using a Level 4 token;

- All Level 4 requirements unrelated to non-repudiation are satisfied.

All Level 1-3 requirements for the protection of assertion data remain in force at Level 4.

## 10. References

This section lists references that are current versions at the time of publication. Subsequent versions of NIST publications (i.e., Federal Information Processing Standards and Special Publications) are also acceptable.

### *10.1. General References*

[DOJ 2000] *Guide to Federal Agencies on Implementing Electronic Processes* (November 2000), available at: http://www.usdoj.gov/criminal/cybercrime/ecommerce.html

[GSA ESIG] *Use of Electronic Signatures in Federal Organization Transactions* (2011), available at: http://www.gsa.gov/

[FISMA] *Federal Information Security Management Act*, available at: http://csrc.nist.gov/drivers/documents/FISMA-final.pdf

[OMB M-04-04] OMB Memorandum M-04-04, *E-Authentication Guidance for Federal agencies*, December 16, 2003, available at: http://www.whitehouse.gov/omb/memoranda/fy04/m04-04.pdf

[OMB M-03-22] OMB Memorandum M-03-22, *OMB Guidance for Implementing the Privacy Provisions of the E-Government Act of 2002*, September 26, 2003 available at: http://www.whitehouse.gov/omb/memoranda/m03-22.html.

[KERB] Neuman, C., and T. Ts'o, *Kerberos: An Authentication Service for Computer Networks*, IEEE Communications, vol. 32, no.9, 1994.

[RFC 4120] IETF, RFC 4120, The Kerberos Network Authentication Service (V5), July 2005, available at http://www.ietf.org/rfc/rfc4120.txt

[RFC 1939] IETF, RFC 1939, *Post Office Protocol,* Version 3, May 1996, available at: http://www.ietf.org/rfc/rfc1939.txt

[RFC 2246] IETF, RFC 2246, *The TLS Protocol,* Version 1.0. January 1999, available at: http://www.ietf.org/rfc/rfc2246.txt

[RFC 5280] IETF, RFC 5280, *Internet X.509 Public Key Infrastructure Certificate and CRL Profile*, available at: http://www.ietf.org/rfc/rfc5280.txt

[RFC 3546] IETF, RFC 3546, *Transport Layer Security (TLS) Extensions*, June 2003, available at: http://www.ietf.org/rfc/rfc3546.txt

[RFC 5246] IETF, RFC 5246, *The Transport Layer Security (TLS) Protocol,* Version 1.2, August 2008, available at http://tools.ietf.org/html/rfc5246

[RFC 1760]   IETF, RFC 1760, *The S/KEY One-Time Password System*, February 1995, available at: http://www.ietf.org/rfc/rfc1760.txt

[ICAM]   National Security Systems and Identity, Credential and Access Management Sub-Committee Focus Group, Federal CIO Council, *ICAM Lexicon*, Version 0.5, March 2011.

[ISPKI]   ITU-T Recommendation X.509 | ISO / IEC 9594-8: "Information Technology - Open Systems Interconnection - The Directory: Public-Key and Attribute Certificate Frameworks."

[SAML]   OASIS, SAML, "Security Assertion Markup Language 2.0," v2.0, March 2005, available at http://www.oasis-open.org/standards#samlv2.0

## 10.2. NIST Special Publications

NIST 800 Series Special Publications are available at: http://csrc.nist.gov/publications/nistpubs/index.html. The following publications may be of particular interest to those implementing systems of applications requiring e-authentication.

[SP 800-30]   NIST Special Publication 800-30, *Risk Management Guide for Information Technology Systems*, July 2002.

[SP 800-32]   NIST Special Publication, 800-32, *Introduction to Public Key Technology and the Federal PKI Infrastructure*, February 2001.

[SP 800-33]   NIST Special Publication 800-33, *Underlying Technical Models for Information Technology Security*, December 2001.

[SP 800-37]   NIST Special Publication 800-37, Revision1, *Guide for Applying the Risk Management Framework to Federal Information Systems*, February 2010.

[SP 800-40]   NIST Special Publication 800-40, Version 2.0, *Creating a Patch and Vulnerability Management Program,*, November 2005.

[SP 800-41]   NIST Special Publication 800-41, Revision 1, *Guidelines on Firewalls and Firewall Policy*, September 2009.

[SP 800-43]   NIST Special Publication 800-43, *Guide to Securing Windows 2000 Professional*, November 2002.

[SP 800-44]   NIST Special Publication 800-44, Version 2,*Guidelines on Securing Public Web Servers*, September 2007.

[SP 800-47]   NIST Special Publication 800-47, *Security Guide for Interconnecting Information Technology Systems*, September 2002.

[SP 800-52]   NIST Special Publication 800-52, *Guidelines for the Selection and Use of Transport Layer Security Implementations*, June 2005.

[SP 800-53]     NIST Special Publication 800-53, Revision 3, *Recommended Security Controls for Federal Information Systems and Organizations*, August 2009 and Errata as of May 2010.

[SP 800-53A]    NIST Special Publication 800-53A, Revision 1, *Guide for Assessing the Security Controls in Federal Information Systems and Organizations, Building Effective Security Assessment Plans*, June 2010.

[SP 800-57]     NIST Special Publication 800-57, Revision 2, *Recommendation for Key Management – Part 1: General*, March 2007.

[SP 800-94]     NIST Special Publication, 800-94, *Guide to Intrusion Detection and Prevention Systems (IDPS)*, February 2007.

[SP 800-115]    NIST Special Publication 800-115, *Technical Guide to Information Security Testing and Assessment*, September 2008.

## 10.3. Federal Information Processing Standards

FIPS can be found at: http://csrc.nist.gov/publications/fips/

[FIPS 140-2]    Federal Information Processing Standard Publication 140-2, *Security Requirements for Cryptographic Modules*, NIST, May 25, 2001.

[FIPS 180-2]    Federal Information Processing Standard Publication 180-2, *Secure Hash Standard (SHS)*, NIST, August 2002.

[FIPS186-2]     Federal Information Processing Standard Publication 186-2, *Digital Signature Standard (DSS)*, NIST, June 2000.

[FIPS 197]      Federal Information Processing Standard Publication 197, *Advanced Encryption Standard (AES)*, NIST, November 2001.

[FIPS 199]      *Standards for Security Categorization of Federal Information and Information Systems* (February 2004), available at: http://csrc.nist.gov/publications/fips/fips199/FIPS-PUB-199-final.pdf

[FIPS 201]      *Personal Identity Verification (PIV) of Federal Employees and Contractors* (March 2006), available at: http://csrc.nist.gov/publications/fips/fips201-1/FIPS-201-1-chng1.pdf

## 10.4. Certificate Policies

These certificate policies can be found at: http://www.cio.gov/fpkipa/

[FBCA1]         *X.509 Certificate Policy For The Federal Bridge Certification Authority (FBCA)*, version 2.1 January 12, 2006. Available at: http://www.cio.gov/fpkipa/documents/FBCA_CP_RFC3647.pdf

[FBCA2]  *Citizen & Commerce Certificate Policy,* Version 1.0 December 3, 2002. Available at:
http://www.cio.gov/fpkipa/documents/citizen_commerce_cpv1.pdf

[FBCA3]  *X.509 Certificate Policy for the Common Policy Framework*, Version 2.4 February 15, 2006. Available at:
http://www.cio.gov/fpkipa/documents/CommonPolicy.pdf

# Appendix A: Estimating Entropy and Strength

## *Password Entropy*

Passwords represent a very popular implementation of memorized secret tokens. In this case impersonation of an identity requires only that the impersonator obtain the password. Moreover, the ability of humans to remember long, arbitrary passwords is limited, so passwords are often vulnerable to a variety of attacks including guessing, use of dictionaries of common passwords, and brute force attacks of all possible password combinations. There are a wide variety of password authentication protocols that differ significantly in their vulnerabilities, and many password mechanisms are vulnerable to passive and active network attacks. While some cryptographic password protocols resist nearly all direct network attacks, these techniques are not at present widely used and all password authentication mechanisms are vulnerable to keyboard loggers and observation of the password when it is entered. Experience also shows that users are vulnerable to "social engineering" attacks where they are persuaded to reveal their passwords to unknown parties, who are basically "confidence men."

Claude Shannon coined the use of the term "entropy[33]" in information theory. The concept has many applications to information theory and communications and Shannon also applied it to express the amount of actual information in English text. Shannon says, "The entropy is a statistical parameter which measures in a certain sense, how much information is produced on the average for each letter of a text in the language. If the language is translated into binary digits (0 or 1) in the most efficient way, the entropy H is the average number of binary digits required per letter of the original language."[34]

Entropy in this sense is at most only loosely related to the use of the term in thermodynamics. A mathematical definition of entropy in terms of the probability distribution function is:

$$H(X) := -\sum_{x} P(X = x) \log_2 P(X = x)$$

where $P(X=x)$ is the probability that the variable $X$ has the value $x$.

Shannon was interested in strings of ordinary English text and how many bits it would take to code them in the most efficient way possible. Since Shannon coined the term, "entropy" has been used in cryptography as a measure of the difficulty in guessing or determining a password or a key. Clearly the strongest key or password of a particular size is a truly random selection, and clearly, on average such a selection cannot be compressed. However it is far from clear that compression is the best measure for the strength of keys and passwords, and cryptographers have derived a number of alternative

---

[33] C. E. Shannon, "A mathematical Theory of Communication," *Bell System Technical Journal*, v. 27, pp. 379-423, 623-656, July, October 1948, see http://cm.bell-labs.com/cm/ms/what/shannonday/paper.html
[34] C. E. Shannon, "Prediction and Entropy of Printed English", *Bell System Technical Journal*, v.30, n. 1, 1951, pp. 50-64.

forms or definitions of entropy, including "guessing entropy" and "min-entropy." As applied to a distribution of passwords the guessing entropy is, roughly speaking, an estimate of the average amount of work required to guess the password of a selected user, and the min-entropy is a measure of the difficulty of guessing the easiest single password to guess in the population.

If we had a good knowledge of the frequency distribution of passwords chosen under a particular set of rules, then it would be straightforward to determine either the guessing entropy or the min-entropy of any password. An Attacker who knew the password distribution would find the password of a chosen user by first trying the most probable password for that chosen username, then the second most probable password for that username and so on in decreasing order of probability until the Attacker found the password that worked with the chosen username. The average for all passwords would be the guessing entropy. The Attacker who is content to find the password of any user would follow a somewhat different strategy, he would try the most probable password with every username, then the second most probable password with every username, until he found the first "hit." This corresponds to the min-entropy.

Unfortunately, we do not have much data on the passwords users choose under particular rules, and much of what we do know is found empirically by "cracking" passwords, that is by system administrators applying massive dictionary attacks to the files of hashed passwords (in most systems no plaintext copy of the password is kept) on their systems. NIST would like to obtain more data on the passwords users actually choose, but, where they have the data, system administrators are understandably reluctant to reveal password data to others. Empirical and anecdotal data suggest that many users choose very easily guessed passwords, where the system will allow them to do so.

## A.1   Randomly Selected Passwords

As we use the term here, "entropy" denotes the uncertainty in the value of a password. Entropy of passwords is conventionally expressed in bits. If a password of $k$ bits is chosen at random there are $2^k$ possible values and the password is said to have $k$ bits of entropy. If a password of length $l$ characters is chosen at random from an alphabet of $b$ characters (for example the 94 printable ISO characters on a typical keyboard) then the entropy of the password is $b^l$ (for example if a password composed of 8 characters from the alphabet of 94 printable ISO characters the entropy is $94^8 \approx 6.09 \times 10^{15}$ – this is about $2^{52}$, so such a password is said to have about 52 bits of entropy). For randomly chosen passwords, guessing entropy, min-entropy, and Shannon entropy are all the same value. The general formula for entropy, $H$ is given by:

$$H = \log_2 (b^l)$$

Table A.1 gives the entropy versus length for a randomly generated password chosen from the standard 94 keyboard characters (not including the space). Calculation of randomly selected passwords from other alphabets is straightforward.

## *A.2 User Selected Passwords*

It is much more difficult to estimate the entropy in passwords that users choose for themselves, because they are not chosen at random and they will not have a uniform random distribution. Passwords chosen by users probably roughly reflect the patterns and character frequency distributions of ordinary English text, and are chosen by users so that they can remember them. Experience teaches us that many users, left to choose their own passwords will choose passwords that are easily guessed and even fairly short dictionaries of a few thousand commonly chosen passwords, when they are compared to actual user chosen passwords, succeed in "cracking" a large share of those passwords.

### A.2.1 Guessing Entropy Estimate

Guessing entropy is arguably the most critical measure of the strength of a password system, since it largely determines the resistance to targeted, online password guessing attacks.

In these guidelines, we have chosen to use Shannon's estimate of the entropy in ordinary English text as the starting point to estimate the entropy of user-selected passwords. It is a big assumption that passwords are quite similar to other English text, and it would be better if we had a large body of actual user selected passwords, selected under different composition rules, to work from, but we have no such resource, and it is at least plausible to use Shannon's work for a "ballpark" estimate. Readers are cautioned against interpreting the following rules as anything more than a very rough rule of thumb method to be used for the purposes of e-authentication.

Shannon conducted experiments where he gave people strings of English text and asked them to guess the next character in the string. From this he estimated the entropy of each successive character. He used a 27-character alphabet, the ordinary English lower case letters plus the space.

In the following discussion we assume that passwords are user selected from the normal keyboard alphabet of 94 printable characters, and are at least 6-characters long. Since Shannon used a 27 character alphabet it may seem that the entropy of user selected passwords would be much larger, however the assumption here is that users will choose passwords that are almost entirely lower case letters, unless forced to do otherwise, and that rules that force them to include capital letters or non-alphabetic characters will generally be satisfied in the simplest and most predictable manner, often by putting a capital letter at the start (as we do in ordinary English) and punctuation or special characters at the end, or by some simple substitution, such as $ for the letter "s." Moreover rules that force passwords to appear to be highly random will be counterproductive because they will make the passwords hard to remember. Users will then write the passwords down and keep them in a convenient (that is insecure) place, such as pasted on their monitor. Therefore it is reasonable to start from estimates of the entropy of simple English text, assuming only a 27-symbol alphabet.

Shannon observed that, although there is a non-uniform probability distribution of letters, it is comparatively hard to predict the first letter of an English text string, but, given the first letter, it is much easier to guess the second and given the first two the third is easier still, and so on. He estimated the entropy of the first symbol at 4.6 to 4.7 bits, declining to on the order of about 1.5 bits after 8 characters. Very long English strings (for example the collected works of Shakespeare) have been estimated to have as little as .4 bits of entropy per character.[35] Similarly, in a string of words, it is harder to predict the first letter of a word than the following letters, and the first letter carries about 6 times more information than the fifth or later letters[36].

An Attacker attempting to find a password will try the most likely chosen passwords first. Very extensive dictionaries of passwords have been created for this purpose. Because users often choose common words or very simple passwords systems commonly impose rules on password selection in an attempt to prevent the choice of "bad" passwords and improve the resistance of user chosen passwords to such dictionary or rule driven password guessing attacks. For the purposes of these guidelines, we break those rules into two categories:

1. Dictionary tests that test prospective passwords against an "extensive dictionary test" of common words and commonly used passwords, then disallow passwords found in the dictionary. We do not precisely define a dictionary test, since it must be tailored to the password length and rules, but it should prevent selection of passwords that are simple transformations of any one word found in an unabridged English dictionary, and should include at least 50,000 words. There is no intention to prevent selection of long passwords (16 characters or more based on phrases) and no need to impose a dictionary test on such long passwords of 16 characters or more.
2. Composition rules that typically require users to select passwords that include lower case letters, upper case letters, and non-alphabetic symbols (e.g.;: "~!@#$%^&*()_-+={}[]|\:;'<,>.?/1234567890").

Either dictionary tests or composition rules eliminate some passwords and reduce the space that an adversary must test to find a password in a guessing or exhaustion attack. However they can eliminate many obvious choices and therefore we believe that they generally improve the "practical entropy" of passwords, although they reduce the work required for a truly exhaustive attack. The dictionary check requires a dictionary of at least 50,000 legal passwords chosen to exclude commonly selected passwords. Upper case letters in candidate passwords should be converted to lower case before comparison.

Table A.1 provides a rough estimate of the average entropy of user chosen passwords as a function of password length. Estimates are given for user selected passwords drawn from the normal keyboard alphabet that are not subject to further rules, passwords subject to a dictionary check to prevent the use of common words or commonly chosen passwords

---

[35] Thomas Schurmann and Peter Grassberger, "Entropy estimation of symbol sequences," http://arxiv.org/ftp/cond-mat/papers/0203/0203436.pdf
[36] *ibid.*

and passwords subject to both composition rules and a dictionary test. In addition an estimate is provided for passwords or PINs with a ten-digit alphabet. The table also shows the calculated entropy of randomly selected passwords and PINs. The values of Table A.1 should not be taken as accurate estimates of absolute entropy, but they do provide a rough relative estimate of the likely entropy of user chosen passwords, and some basis for setting a standard for password strength.

The logic of the Table A.1 is as follows for user-selected passwords drawn from the full keyboard alphabet:

- The entropy of the first character is taken to be 4 bits;
- The entropy of the next 7 characters are 2 bits per character; this is roughly consistent with Shannon's estimate that "when statistical effects extending over not more than 8 letters are considered the entropy is roughly 2.3 bits per character;"
- For the 9th through the 20th character the entropy is taken to be 1.5 bits per character;
- For characters 21 and above the entropy is taken to be 1 bit per character;
- A "bonus" of 6 bits of entropy is assigned for a composition rule that requires both upper case and non-alphabetic characters. This forces the use of these characters, but in many cases these characters will occur only at the beginning or the end of the password, and it reduces the total search space somewhat, so the benefit is probably modest and nearly independent of the length of the password;
- A bonus of up to 6 bits of entropy is added for an extensive dictionary check. If the Attacker knows the dictionary, he can avoid testing those passwords, and will in any event, be able to guess much of the dictionary, which will, however, be the most likely selected passwords in the absence of a dictionary rule. The assumption is that most of the guessing entropy benefits for a dictionary test accrue to relatively short passwords, because any long password that can be remembered must necessarily be a "pass-phrase" composed of dictionary words, so the bonus declines to zero at 20 characters.

For user selected PINs the assumption of Table A.1 is that such pins are subjected at least to a rule that prevents selection of all the same digit, or runs of digits (e.g., "1234" or "76543"). This column of Table A.1 is at best a very crude estimate, and experience with password crackers suggests, for example, that users will often preferentially select simple number patterns and recent dates, for example their year of birth.

### A.2.2 Min-Entropy Estimates

Experience suggests that a significant share of users will choose passwords that are very easily guessed ("password" may be the most commonly selected password, where it is allowed). Suppose, for example, that one user in 1,000 chooses one of the 2 most common passwords, in a system that allows a user 3 tries before locking a password. An Attacker with a list of user names, who knows the two most commonly chosen passwords can use an automated attack to try those 2 passwords with each user name, and can expect to find at least one password about half the time by trying 700 usernames with those two passwords. Clearly this is a practical attack if the only goal is to get access to the system, rather than to impersonate a single selected user. This is usually too dangerous a possibility to ignore.

We know of no accurate general way to estimate the actual min-entropy of user chosen passwords, without examining in detail the passwords that users actually select under the rules of the password system, however it is reasonable to argue that testing user chosen passwords against a sizable dictionary of otherwise commonly chosen legal passwords, and disallowing matches, will raise the min-entropy of a password. A dictionary test is specified here that is intended to ensure at least 10 bits of min-entropy. That test is:

- Upper case letters in passwords are converted to entirely lower case and compared to a dictionary of at least 50,000 commonly selected otherwise legal passwords and rejected if they match any dictionary entry, and

- Passwords that are detectable permutations of the username are not allowed.

This is estimated to ensure at least 10 bits of min-entropy. Other means may be substituted to ensure at least 10 bits of min-entropy. User chosen passwords of at least 15 characters are assumed to have at least 10 bits of min-entropy. For example a user might be given a short randomly chosen string (two randomly chosen characters from a 94-bit alphabet have about 13 bits of entropy). A password, for example, might combine short system selected random elements, to ensure 10 bits of min-entropy, with a longer user-chosen password.

## A.3 Other Types of Passwords

Some password systems require a user to memorize a number of images, such as faces. Users are then typically presented with successive fields of several images (typically 9 at a time), each of which contains one of the memorized images. Each selection represents approximately 3.17 bits of entropy. If such a system used five rounds of memorized images, then the entropy of system would be approximately 16 bits. Since this is randomly selected password the guessing entropy and min-entropy are both the same value.

It is possible to combine randomly chosen and user chosen elements into a single composite password. For example a user might be given a short randomly selected value to ensure min-entropy to use in combination with a user chosen password string. The random component might be images or a character string.

Table A.1 – Estimated Password Guessing Entropy in bits vs. Password Length

| | User Chosen | | | Randomly Chosen | |
| --- | --- | --- | --- | --- | --- |
| | 94 Character Alphabet | | | 10 char. alphabet | 94 char alphabet |
| Length Char. | No Checks | Dictionary Rule | Dict. & Composition Rule | | |
| 1 | 4 | - | - | 3 | 3.3 | 6.6 |
| 2 | 6 | - | - | 5 | 6.7 | 13.2 |
| 3 | 8 | - | - | 7 | 10.0 | 19.8 |
| 4 | 10 | 14 | 16 | 9 | 13.3 | 26.3 |
| 5 | 12 | 17 | 20 | 10 | 16.7 | 32.9 |
| 6 | 14 | 20 | 23 | 11 | 20.0 | 39.5 |
| 7 | 16 | 22 | 27 | 12 | 23.3 | 46.1 |
| 8 | 18 | 24 | 30 | 13 | 26.6 | 52.7 |
| 10 | 21 | 26 | 32 | 15 | 33.3 | 65.9 |
| 12 | 24 | 28 | 34 | 17 | 40.0 | 79.0 |
| 14 | 27 | 30 | 36 | 19 | 46.6 | 92.2 |
| 16 | 30 | 32 | 38 | 21 | 53.3 | 105.4 |
| 18 | 33 | 34 | 40 | 23 | 59.9 | 118.5 |
| 20 | 36 | 36 | 42 | 25 | 66.6 | 131.7 |
| 22 | 38 | 38 | 44 | 27 | 73.3 | 144.7 |
| 24 | 40 | 40 | 46 | 29 | 79.9 | 158.0 |
| 30 | 46 | 46 | 52 | 35 | 99.9 | 197.2 |
| 40 | 56 | 56 | 62 | 45 | 133.2 | 263.4 |

Figure A.1 - *Estimated User Selected Password Entropy vs. Length*

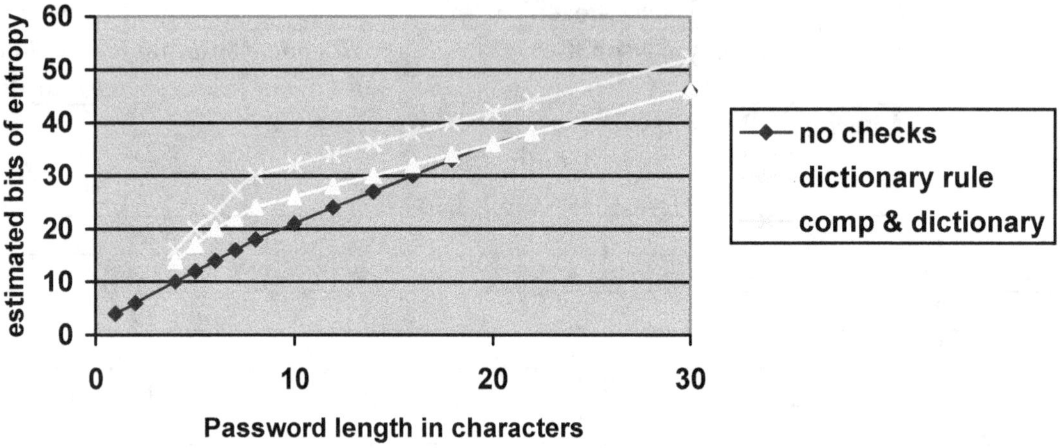

# Appendix B: Mapping of Federal PKI Certificate Policies to E-authentication Assurance Levels

Agencies are, in general, issuing certificates under the policies specified in the Common Policy Framework [FBCA3] to satisfy FIPS 201. Organizations outside the US Government have begun issuing credentials under a parallel set of policies and requirements known collectively as PIV Interoperabiltiy specifications (PIV-I). Agencies that were early adopters of PKI technology, and organizations outside the Federal government, issue PKI certificates under organization specific policies instead of the Common Policy Framework. The primary mechanism for evaluating the assurance provided by public key certificates issued under organization specific policies is the policy mapping of the Federal Policy Authority to the Federal Bridge CA policies. These policies include the Rudimentary, Basic, Medium, Medium-HW, and High assurance policies specified in [FBCA1] and the Citizen and Commerce class policy specified in [FBCA2].

These policies incorporate all aspects of the credential lifecycle, often in greater detail than specified here. These policies also include security controls (e.g., multi-party control and system auditing for CSPs) that are outside the scope of this document. However, the FPKI policies are based on work that largely predates this specification, and the security requirements are not always strictly aligned with those specified here. As a result, this appendix provides an overall mapping between FPKI certificate policies and the e-authentication Levels instead of a strict evaluation of compliance. There are known discrepancies, such as FIPS 201's allowance for pseudonyms on credentials issued to personnel in dangerous jobs, or the ability to issue PIV credentials based on a single federal government issued identity credential. While these discrepancies are recognized, the overall level of assurance provided by these policies is deemed to meet the requirements based on the additional controls.

The table below summarizes how certificates issued under the Common Policy Framework correspond to the e-authentication assurance levels. Note that the Common Device policy is not listed; this policy supports authentication of a system rather than a person. In addition, table B.1 summarizes how organization specific certificate policies correspond to e-authentication assurance levels. At Level 2 agencies may use certificates issued under policies that have not been mapped by the Federal policy authority, but are determined to meet the Level 2 identify proofing, token and status reporting requirements. (For this evaluation, a strict compliance mapping should be used, rather than the rough mapping used for the FPKI policies.) For Levels 3 and 4, agencies shall depend upon the mappings provided by the Federal PKI.

The Federal PKI has also added two policies, Medium Commercial Best Practices (Medium-CBP) and Medium Hardware Commercial Best Practices (MediumHW-CBP) to support recognition of non-Federal PKIs. In terms of e-authentication levels, the Medium CBP and MediumHW-CBP are equivalent to Medium and Medium-HW, respectively.

Table B.1 – Certificate Policies and the E-authentication Assurance Levels

| Certificate Policy | Selected Policy Components | | | Overall Equivalence |
|---|---|---|---|---|
| | Identity Proofing | Token | Token and Credential Management[37] | |
| Common-Auth<br>PIVI-Auth<br>SHA1-Auth[38] | Meets Level 4 | Meets Level 4 | Meets Level 4 | Meets Level 4 |
| Common-SW | Meets Level 4 | Meets Level 3 | Meets Level 4 | Meets Level 3 |
| Common-HW<br>PIVI-HW<br>SHA1-HW[38] | Meets Level 4 | Meets Level 4 | Meets Level 4 | Meets Level 4 |
| Common-High | Meets Level 4 | Meets Level 4 | Meets Level 4 | Meets Level 4 |
| FBCA Basic[39] | Meets Level 3 | Meets Level 3 | Meets Level 3 | Meets Level 3 |
| FBCA Medium[39] | Meets Level 4 | Meets Level 3 | Meets Level 4 | Meets Level 3 |
| FBCA Medium-HW[39] | Meets Level 4 | Meets Level 4 | Meets Level 4 | Meets Level 4 |
| FBCA High[39] | Meets Level 4 | Meets Level 4 | Meets Level 4 | Meets Level 4 |
| Common-cardAuth<br>PIVI-cardAuth<br>SHA1-cardAuth[38] | Meets Level 4 | Meets Level 2 | Meets Level 4 | Meets Level 2 |

---

[37] The key component in token and credential management is the credential status mechanism.
[38] The SHA1 policies have been deprecated and will not be acceptable after December 31, 2013.
[39] These policies are not asserted in the user certificates, but equivalence is established through policy mapping at the Federal Bridge CA.

www.ingramcontent.com/pod-product-compliance
Lightning Source LLC
Chambersburg PA
CBHW081728170526
45167CB00009B/3737